Forest and Plant Bioproducts Division 2022

Held at the 2022 AIChE Annual Meeting

Phoenix, Arizona, USA
13-18 November 2022

ISBN: 978-1-7138-7886-5

Forest and Plant Bioproducts Division 2022

Held at the 2022 AIChE Annual Meeting

Phoenix, Arizona, USA
13-18 November 2022

AIChE
The Global Home of Chemical Engineers

Printed from e-media with permission by:

Curran Associates, Inc.
57 Morehouse Lane
Red Hook, NY 12571

Some format issues inherent in the e-media version may also appear in this print version.

Printed with permission by Curran Associates, Inc. (2023)

For permission requests, please contact AIChE
at the address below.

AIChE
120 Wall Street, FL 23
New York, NY 10005-4020

Phone: (800) 242-4363
Fax: (203) 775-5177

www.aiche.org

Additional copies of this publication are available from:

Curran Associates, Inc.
57 Morehouse Lane
Red Hook, NY 12571 USA
Phone: 845-758-0400
Fax: 845-758-2633
Email: curran@proceedings.com
Web: www.proceedings.com

TABLE OF CONTENTS

ADVANCED BIOMASS CONVERSION TECHNOLOGIES

ADVANCED SEPARATIONS PROCESSES IN BIOPROCESSING AND BIOMATERIALS

ADVANCES IN LIGNIN DEGRADATION STRATEGIES

LIGNIN FOR SUSTAINABLE INDUSTRIAL USES

POSTER SESSION: NOVEL PRODUCTS FROM FOREST AND PLANT BIOMASS

PROCESS INTENSIFICATION IN BIOREFINERIES

THERMOCHEMICAL CONVERSION OF BIOMASS

VALUE-ADDED USES OF INDUSTRIAL COPRODUCTS AND NATURAL FIBRES IN SUSTAINABLE USES

WASTE VALORIZATION

ANDREW CHASE AWARD II (INVITED TALKS)

CHEMICAL MODIFICATIONS AND PROCESSING OF BIOMATERIALS II

451a Low Temperature Hydrothermal Liquefaction of a Mixture of Lignin and Hemicellulose Derived Product Mixture

Sampath Gunukula, Sampath Karunarathne, Clayton Wheeler

About 60% of Maine's pulp and paper mills closed between 2000 and 2017. Thousands of workers were laid off during this period. Forest owners have also been impacted by decreased demand wood. These problems can be overcome by creating additional revenue streams for pulp and papermills by upgrading currently underutilized lignin and hemicelluloses. In this study, we perform low temperature (250 °C) hydrothermal liquefaction of a mixture of lignin and hemicellulose-derived xylose to transportation fuel intermediates. We further convert hemicellulose to fatty acids using microbial strains and study the hydrothermal liquefaction of a mixture of lignin and hemicellulose derived fatty acids. The effectiveness of three different base catalysts (sodium hydroxide, potassium hydroxide, and potassium carbonate) will be shown for the low temperature hydrothermal liquefaction of lignin/hemicellulose mixture and lignin/fatty acids mixture. The resulting hydrothermal liquids are characterized using the gas chromatography-mass spectroscopy (GC-MS) method. These GC-MS characterization results will be used to determine the need for upgrading hemicelluloses to fatty acids prior to the hydrothermal liquefaction, suitable base catalyst for the low temperature hydrothermal liquefaction, and screening further upgrading hydrothermal liquefaction liquids to transportation fuels.

451b Microwave Heated Pretreatment of Agricultural Residues with Deep Eutectic Solvents to Isolate Lignin-Carbohydrate Complex for Renewable Fuel and Material Applications

Sanphawat Phromphithak, Thossaporn Onsree, Jochen Lauterbach, Nakorn Tippayawong

Agricultural residues can be used as renewable and sustainable materials in downstream production of biochemicals and biofuels. Generally, agricultural residues contain lignin-carbohydrate complex, making conversion of agricultural residues to biochemicals and/or biofuels challenging. Fractionation of lignocellulosic biomass offers a possible solution for specific applications. Deep eutectic solvents (DESs), a combination of two or more chemical components through hydrogen bonding, can be promisingly applied to dissolve and isolate lignocellulose compositions. Microwave heating technology has been intensely developed to provide rapid, low energy, and high efficiency. It may be applied to biomass pretreatment. So far, there have only been few studies on microwave heated DES pretreatment of agricultural residues. In this work, pretreatment of corn stover with DESs was investigated in a microwave heating reactor with a fixed biomass loading of 10 wt.% at varying reaction temperature (60-150 °C) and residence time (2-10 min). The DESs were prepared form choline derivative and glycerol with different ratios of 1:3-12. The resulting solid product is considered as cellulose-rich materials (CRMs). The CRM was analyzed by thermogravimetry (TG), scanning electron microscope (SEM), Fourier-transform infrared spectroscopy (FTIR), and X-ray diffraction (XRD). Compared to conventional heating, the microwave-assisted heating provided a significantly better efficiency for eliminating non-cellulosic compounds from the raw feedstock. SEM images showed that CRM fiber bundles were destroyed into small fragments with micro-sizes. FTIR results indicated a reduced OH peak intensity of lignin and disappearing ester linkage between lignin and hemicellulose. XRD illustrated that cellulose crystallinity was increased by 30 to over 70%. Using nuclear magnetic resonance spectroscopy, the microwave heated pretreatment was shown to affect more interactions between the DES and lignin structure, improving efficiency of breaking down lignocellulosic biomass structures. Our findings are useful for biomass utilization and applications in generating bioenergy, fuels, and chemicals.

451c Improving Bamboo's Fuel and Storage Properties with a Net Energy Export through Torrefaction Paired with Catalytic Oxidation

Nepu Saha, Eric Fillerup, Brad Thomas, Corey Pilgrim, Thomas Causer, Dan Herren, Jordan Klinger

Using torrefied char, or biocoal, as solid fuel provides an opportunity to introduce a sustainable feedstock into the energy market. The goals of this study were to investigate how torrefaction improves the energy content and the grindability of a Malaysian bamboo along with understanding the potential for integrated energy recovery from torrefaction gases. The feedstock was torrefied at 250–290 °C for 1 h and the combustion characteristics and grindability of the solid products along with the composition of torrefied gas species were measured. The results showed a beneficial increase in elemental carbon increased from 47 to 63 wt% at 290 °C torrefaction, reflecting an increase in higher heating value from 17.8 to 25.6 MJ/kg. The combustion behavior of all the products appeared to show three distinct combustion zones; however, torrefaction severity leading to higher combustion temperature and time as the fixed carbon augmented with the increase of torrefaction temperature. This increase in severity also lead to more friable and grindable material, and the 290 °C condition required a factor of 2.7 less hold-up time in the mill compared to the raw bamboo, and a factor of 8.5 less energy (938 and 111 kWh/tonne respectively). Through analysis of the gas and volatile formation, a case study showed that catalytic oxidation can convert nearly 100% of the embodied chemical energy into usable thermal energy. These experimental findings were scaled to a 100,000 tonne/y capacity torrefaction plant and in the moderate case of 270 °C operating temperature, the plant has 1.25 GJ/tonne excess energy beyond what the process needs.

451d Computational Fluid Dynamic Simulations for Woodchips Chemical Looping Gasification in a Bubbling Fluidized Bed Fuel Reactor

Hessamedin Naeimi, Afsaneh Khajeh, Lijun Wang, Abolghasem Shahbazi

Using the gasification process, in a fluidized bed reactor, in conjunction with chemical looping technology has been found to be one of the most promising ways for biomass conversion called chemical looping gasification (CLG). In this study, a two-dimensional transient computational fluid dynamic (CFD) model is developed to simulate hydrodynamics, the homogeneous and heterogeneous reaction rates taking place in chemical looping gasification of woodchips as fuel and iron oxide (Fe_2O_3) as oxygen carrier (OC) in a bubbling fluidized bed fuel reactor. All three phases including two solid phases and one gas phase are treated using a two-fluid model (TFM) scheme based on the Eulerian-Eulerian multiphase approach. First, the predicted results by CFD are compared with experimental data where a good agreement is achieved. Then, the impact of operating conditions; temperature, steam to biomass, and oxygen content of OC to biomass; on the outlet gas composition distribution and chemical looping gasification (CLG) performance are analyzed. The simulation results demonstrate that these operating conditions have a considerable effect on the gas yield and CLG efficiency. It is found that higher temperature increases the syngas yield, gasification, and carbon conversion efficiency due to favoring endothermic reactions at escalated temperature while the H_2/CO concentration ratio decreases because of higher reactivity of H_2 with OC. Also, higher steam to biomass ratio promotes the H_2 concentration as it enhances carbon-steam gasification reaction and water-gas shift reaction. Raising the feeding rate of OC leads to a marked drop in H2 and CO relative concentrations but a higher carbon efficiency due to a relatively higher OC-Fuel ratio.

318c Microcrystalline Cellulose from Soybean Hull As an Excipient for Solid Dosage Forms: Powder Characterization and Tableting Properties

Navid Etebari Alamdari, Burak Aksoy, Zhihua Jiang, Jayachandra B. Ramapuram

Soybean is one of the most valuable crops being planted and harvested worldwide. The USA is the leading country in soy production (around 124 million metric tons in 2018-19), followed by Brazil, Argentine, China, India, Paraguay, and Canada. The main products of the soy industry are soy oil and soy protein with many food, feed, and cosmetic applications. The seed coat of soybeans, known as soybean hull (SBH), is the major by-product of the soybean industry and represents 8-10% of the whole composition of the grain. Despite the large amount generated, SBHs have received minimal attention as low-cost raw materials, and their applications are limited to animal feed and dietary fiber, while most of them are still left to waste. Therefore, there is a strong need to create higher value-added products from this agro-industrial waste.

Microcrystalline cellulose (MCC) is one of the most widely used binders in the formulation of oral suspensions and lotions, oral tablets, and capsules due to its chemical purity, non-toxicity, biodegradability, superior disintegrating property, and greater flowability. According to a market study conducted by Future Market Insights (FMI), the market for MCC will grow at a steady annual rate of 4.7%, topping US$ 818 million in 2028. Among all the markets mentioned above, the pharmaceutical industry dominates the global MCC demand with about 40 percent market share, most of which can be attributed to its role as the direct compression (DC) binder in oral tablet manufacturing.

In this study, we have produced SBH-based MCC as an alternative to wood-based commercial MCC (Avicel®). The processing stages include acid pretreatment followed by filtering and washing, alkali pulping, and chlorine dioxide bleaching. The filtrate from the acid treatment stage can be used in producing biopolymer-based (rich in protein and pectin) coatings and films. This integrated approach will produce a by-product while decreasing the amount of effluent produced. SBH-based powders were tested and compared with Avicel MCC powders for their thermal properties (TGA), structural and morphological properties (XRD and SEM), chemical composition (HPLC), density, and flowability (tap density tester, pycnometer, and angle of repose), and surface area (BET). The powders were then incorporated as the main component (60%) of excipient-only tablets comprised of lactose, sodium starch glycolate, and magnesium stearate. The quality and properties of the produced SBH MCC tablets (hardness, tensile strength, and disintegration) were tested and compared to the Avicel MCC tablets. Based on the results, we concluded that the SBH-based MCC can be a potential alternative for industrial scale tablet production.

318b Empowering Carbon Negative Bio-Chemicals with Advanced Electrochemical Separations

Marcel Gausmann, Christian Schröder, Andreas Jupke

Succinic acid has the potential to become a central platform molecule for renewable and bio-based polyesters. [1] The biomass based polyesters synthesized from succinic acid can substitute petrochemically made polymers while having 34-40% less climate impact. [2] A key feature of succinic acid is that the prokaryotic strains like *Basfia succiniciproducens* or *Actinobacillus succinogenes* have the ability to utilize CO2 as a secondary carbon substrate under anaerobic conditions. This brings the opportunity to make the industrial succinic acid production a carbon sink and provide carbon capture capacity that creates synergies with other industries like bioethanol or cement.

However, efficient downstream technologies with low process carbon footprints are necessary to preserve the carbon capture capacity of bio-succinic acid production. In contrast, most techno economic studies recommend to focus on the development of aerobic and low pH tolerant yeast strains. [4] A low pH fermentation reduces the complexity of the downstream processing at the expense of CO2 emissions from the up-stream. The newly emerging electrochemical separation technologies [5–9] allow a waste free separation of succinic acid. We compared the state of the art downstream technologies for succinic acid purification with in situ separation by electrochemical pH swing extraction. First, we investigate the thermodynamic limits of the microbial metabolism in prokaryotic and eukaryotic host cells and identify a tradeoff between microbial energy demand for product export under low pH conditions and high carbon conversion yields which require a neutral pH in the fermentation. Second, we quantify the effect of these limits on a process scale by comparing different conceptual process designs for succinic acid purification in an extension of a previous techno economic assessment. The electrochemical pH swing separation revealed the potential to maintain the negative carbon emission balance for succinic acid production, while also reducing the water consumption and waste footprint in comparison to the state of the art technologies.[10] Finally, we implemented a dynamic process model in MATLAB to validate the feasibility of in situ product removal by electrochemical pH swing extraction and electrochemical pH control of a dual substrate (glycerol and CO2) succinic acid fermentation.[11] In a sensitivity analysis we compared the reaction kinetic of the fermentation with the kinetics of a pH swing electrolysis with integrated reactive extraction of succinic acid by tertiary amines. The sensitivity study shows that the electrolysis is the rate limiting step for product recovery but significantly faster than the production rate of the fermentation, thus making electrochemical pH-control feasible. We applied the dynamic model to investigate the potential to operate the succinic acid production with a closed base recycle loop. The results show that if concentrated feedstock is available, the consumption of pH control agent in the fermentation is reduced by 88% and the fresh water demand can be reduced by 99% compared to the open loop process.[11] The electrification of succinic acid downstream processing, when operated with renewable electricity, turns bio-succinic acid into a carbons sink for other industries and thereby provides a promising perspective for industrial carbon integration.

Literature Cited

1. Hsu K-H, Chen C-W, Wang L-Y, Chan H-W, He C-L, Cho C-J, Rwei S-P, Kuo C-C. Bio-based thermoplastic poly(butylene succinate-co-propylene succinate) copolyesters: effect of glycerol on thermal and mechanical properties. *Soft matter.* 2019;15(47):9710–9720.

2. Patel MK, Crank M, Dornburg V, Hermann B, Roes L. *Medium and Long-term Opportunities and Risks of the Biotechnological Production of Bulk Chemicals from Renewable Resources.*, 2006.

3. Dessie W, Wang Z, Luo X, Wang M, Qin Z. Insights on the Advancements of In Silico Metabolic Studies of Succinic Acid Producing Microorganisms: A Review with Emphasis on Actinobacillus succinogenes. *Fermentation.* 2021;7(4):220.

4. Dickson R, Mancini E, Garg N, Woodley JM, Gernaey KV, Pinelo M, Liu J, Mansouri SS. Sustainable bio-succinic acid production: superstructure optimization, techno-economic, and lifecycle assessment. *Energy Environ. Sci.* 2021;14(6):3542–3558.

5. Handojo L, Wardani AK, Regina D, Bella C, Kresnowati MTAP, Wenten IG. Electro-membrane processes for organic acid recovery. *RSC Adv.* 2019;9(14):7854–7869.

6. Pateraki C, Andersen SJ, Ladakis D, Koutinas A, Rabaey K. Direct electrochemical extraction increases microbial succinic acid production from spent sulphite liquor. *Green Chem.* 2019;21(9):2401–2411.

7. Gausmann M, Kocks C, Doeker M, Eggert A, Maßmann T, Jupke A. Recovery of succinic acid by integrated multi-phase electrochemical pH-shift extraction and crystallization. *Separation and Purification Technology.* 2020;240(240):116489.

8. Gausmann M, Kocks C, Pastoors J, Büchs J, Wierckx N, Jupke A. Electrochemical pH-T-Swing Separation of Itaconic Acid for Zero Salt Waste Downstream Processing. *ACS Sustainable Chem. Eng.* 2021;9(28):9336–9347.

9. Gausmann M, Gössi A, Bertram F, Riedl W, Schuur B, Jupke A. Electrochemical membrane-assisted pH-swing extraction and back-extraction of lactic acid. *Separation and Purification Technology.* 2022;289(3):120702.

10. Gausmann M, Jupke A. Techno economic assessment of downstream technologies for carbon negative bio-succinic acid production. *Manuscript in preparation.* 2022.

11. Gausmann M, Kiefel R, Jupke A. Dynamic modeling of electrochemical pH swing extraction for in situ separation of succinic acid reveals biorefinery integration potential. *Manuscript in submission.* 2022.

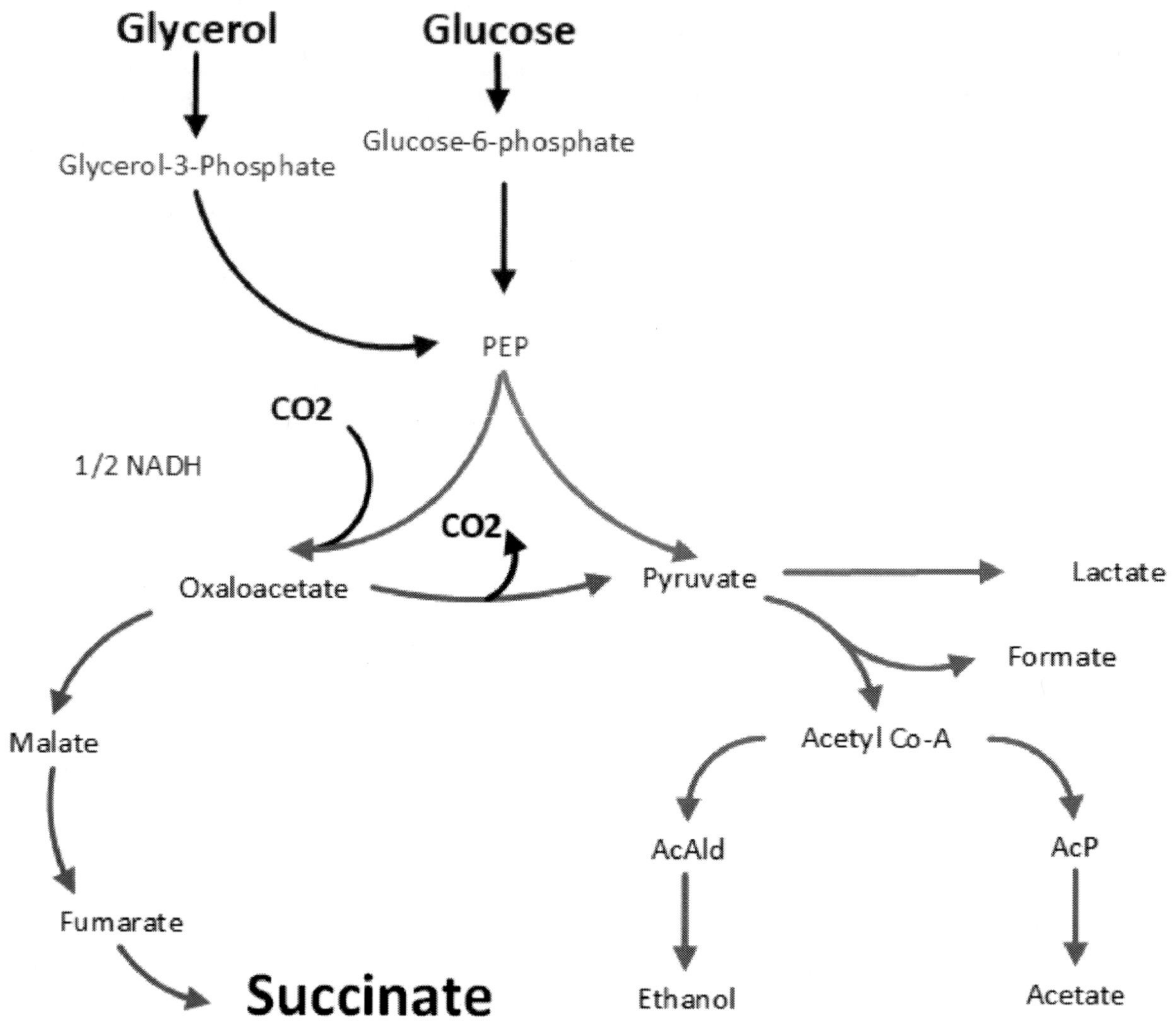
Glycerol
Glucose
Glycerol-3-Phosphate
Glucose-6-phosphate
PEP
CO2
1/2 NADH
CO2
Oxaloacetate
Pyruvate
Lactate
Formate
Malate
Acetyl Co-A
AcAld
AcP
Fumarate
Succinate
Ethanol
Acetate

318a Robust Separation Process for Producing Pure Glucaric Acid Crystals from Fermentation Broth

Hoon Choi, Nathan Soland, Bonnie L. Buss, Stefan Haugen, Joel Miscall, Eric Tan, Eric Karp

Glucaric acid (GA) is regarded as a top-value added bio-derived compound. GA and its salts have applications in many products including biodegradable detergents, corrosion inhibitors, and polymers precursors or additives. Today, GA is primarily produced via the chemical oxidation of glucose using nitric acid or catalytic oxidation of glucose. These approaches have either an expensive and nonselective or a lower conversion yield and scalability issues. Alternatively, biocatalysis offers high selectivity and conversion yield in mild reaction conditions such that fermentation is promising for producing GA with several green chemistry benefits. However, a major challenge in this bioprocessing is to separate and purify GA from the fermentation broth. That is because GA is dissolved in complex mixtures and, most importantly, GA diacid is readily lactonized into D-glucaro-1,4-lactone, D-glucaro-6,3-lactone, and D-glucaro-1,4:6,3-dilactone in aqueous conditions at low pH. Accordingly, an efficient method for glucaric acid separation, especially its diacid form, is necessary to facilitate its valorization. In this seminar, we present a robust separation process that produces glucaric acid crystals from fermentation broth. The developed process is divided into two-step purification processes: (1) recovering monopotassium glucarate (KGA) from broth and then (2) crystallizing purified GA from KGA through acidification and antisolvent crystallization. Overall, this process was found to separate monopotassium glucarate and glucaric acid with a recovery yield of >99.9 % and 71 % at purities of c.a. 95.6 and 98.3 %, respectively. The overall energy footprint of the developed downstream process will also be discussed based on the post-solvent recovery model.

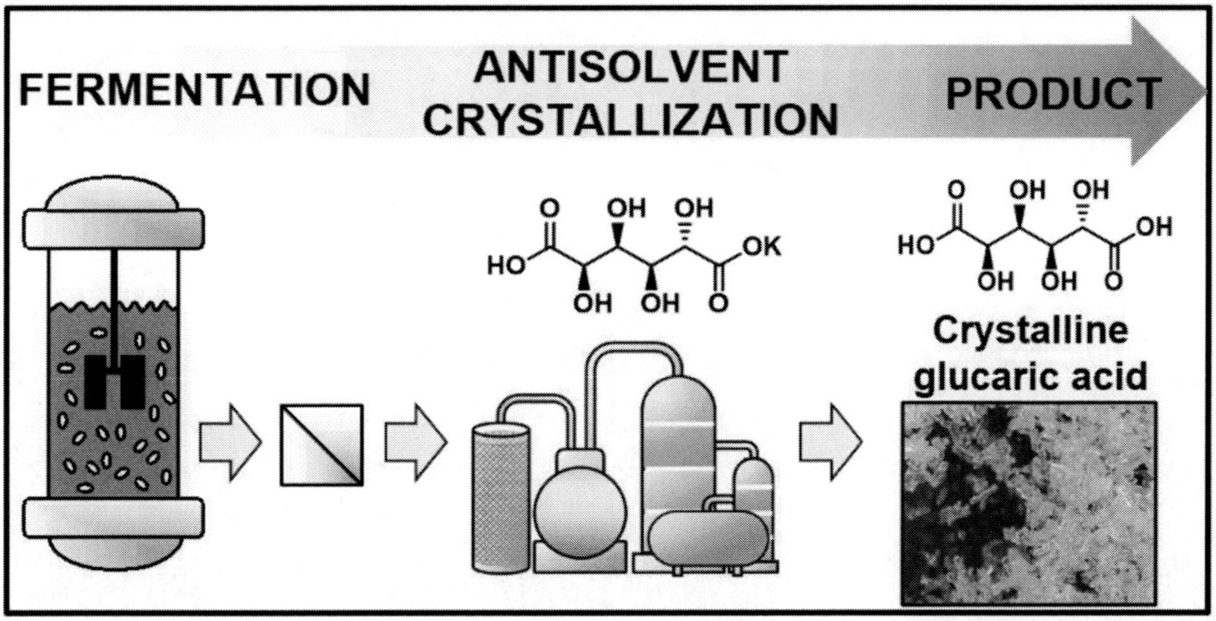

318d Numerical Analysis of Conventional Drying of Paper and Board

Xinyi Li, Huajiang Huang, Sridharan Ramaswamy

Paper and board continue to be an important renewable bioproducts manufacturing sector in the US and the world. Conventional drying processes used in majority of the paper and board are based in conduction heat transfer from steam heated dryer cans and convective heat and mass transfer from air flowing over the surface of the paper as it is being dried. Fundamental understanding of the transport processes involved in conventional drying processes in porous materials like paper is an important endeavor that has attracted much attention. Here we present an approach to model the conduction and convection and associated liquid, gas and energy flows during the drying process. Preliminary results from the model showing the capabilities in predicting some of the key elements of the process are presented. As part of the ongoing effort, we plan to continue to advance the modeling effort and gain detailed understanding of the transport processes. This will help in better understanding of the conventional processes and the development of novel drying processes.

63a A Study for Kinetic Modeling and Experimental Validation of Lignin Fractionation with 4-Phenolsulfonic Acid for Effective Lignocellulosic Biomass Utilization

Juhyeon Kim, Hyun-Kyu Choi, Anqi Ji, Mairui Zhang, Joseph Kwon, Chang Geun Yoo

In terms of sustainability and renewability, biomass utilization has received significant research interest, instead of consuming only petrochemical products, to obtain alternative resources such as energy, aromatic chemicals, etc. [1] However, those biorefineries are still suffering from their low cost-competitiveness, compared to the existing petrorefineries, despite considerable research efforts on effective biomass utilization. To improve the competitiveness of biomass-derived products, it is necessary to maximize the use of major biomass components including cellulose and lignin via effective refinery strategies. Among the major biomass components, lignin is a crucial recalcitrance factor since it acts as either a physical barrier limiting cellulose accessibility to chemicals and/or enzymes, or an inhibitor reducing enzymes and microorganism activities, resulting in poor productivity of the end-products like biofuel [2]. Consequently, effective lignin removal is desired in most biomass utilization strategies. There have been a number of studies for biomass pretreatment/fractionation using acid, alkaline, organic solvent, or hot water [3]. However, these conventional methods had unwanted lignin degradation and modification due to their harsh reaction conditions. Therefore, recently fractionation processes have been developed to yield not only carbohydrates but also high-quality lignin [4].

In this sense, 4-phenolsulfonic acid (PSA) has been considered as an effective fractionation reagent [2]. The amphiphilic property of PSA makes effective lignin fractionation under mild conditions. In our previous study [3], the technical feasibility of PSA in biomass fractionation was conceptionally proven by experiments; however, the development of the systematic model and process optimization is still necessary for maximizing process efficiency. To achieve these objectives, our model was initially developed based on a series of experiments. That is, underlying kinetics of the reactions were firstly examined from the experimental data with different biomass dimensions and reaction temperatures using aspen wood chip as bulk biomass and PSA as a solvent. Afterward, the kinetic Monte Carlo (kMC) algorithm was combined with a governing equation, *i.e.*, mass balance equation of major component (e.g., cellulose, lignin), to model a sophisticated series of phenomena. By using the above-developed multiscale model, one can track not only macroscopic phenomena (*e.g.*, mass/energy transfer) but also microscopic events (*e.g.*, tracing porosity and cell wall thickness); therefore, the biomass properties can extensively be understood throughout the entire process period[5]. Especially, the model includes consideration for delignification and de/repolymerization processes of lignin fragments in the reacting solution to overcome underutilization of the lignin.

Literature cited:

[1] Kim K.H., andamp; Kim C.S. (2018). Recent Efforts to Prevent Undesirable Reactions From Fractionation to Depolymerization of Lignin: Toward Maximizing the Value From Lignin. *Front. Energy Res., 6*, 92.

[2] Yoo C.G., Meng X., Pu Y., andamp; Ragauskas A.J. (2020). The critical role of lignin in lignocellulosic biomass conversion and recent pretreatment strategies: A comprehensive review. *Bioresour. Technol., 301,* 122784.

[3] He D., Wang Y., Yoo C.G., Chen Q.-J., andamp; Yang Q. (2020). The fractionation of woody biomass under mild conditions using bifunctional phenol-4-sulfonic acid as a catalyst and lignin solvent. *Green Chem., 22,* 5414-5422.

[4] Morais A.R.C., da Costa Lopes A.M., andamp; Bogel-Łukasik R. (2015). Carbon Dioxide in Biomass Processing: Contributions to the Green Biorefinery Concept. *Chem. Rev., 115,* 3-27.

[5] Choi H.-K., andamp; Kwon J.S.-I. (2019). Modeling and control of cell wall thickness in batch delignification. *Comput. Chem. Eng. 128*, 512-523.

63b Selective Oxidative Conversion of Acetylated Lignin Towards Aromatic Acids

Wenbo Peng, Zhaohui Tong, Hanxi Bao, Dequan Xiao, Yigui Wang, William 'Joe' Sagues, Helena Hagelin Weaver

A variety of lignin valorization methods have been employed to convert lignin to value-added biochemicals to promote biorefinery profitability. However, a route that can isolate a single type of value-added biochemicals from a mixture of lignin-depolymerized products has become a fascinating challenge. Herein, we propose a new three-step route to depolymerize the industrial lignin into a decent yield of targeted aromatic acids including p-hydroxybenzoic acid, vanillic acid and syringic acid using low-cost GO-UHP catalyst system and under mild conditions (atm, <120 C). In this route, metal-free graphene oxide selectively depolymerizes the acetylation-induced lignin towards aromatic aldehydes/acids. After that, a protective oxidizing upgrading method using UHP as the oxidizer converts the aromatic aldehydes in the lignin depolymerized mixture to targeted aromatic acids with an excellent yield (up to 18%). This route achieves an overall 95% organic oil yield, more importantly, a decent yield of AA can be isolated through a simple base/acid extraction method. Mechanisms related to each step are proposed to elucidate the role of catalyst, solvent and the impurities in the process, and quantum chemistry calculations are also adopted to elucidate the mechanism of this three-step oxidative lignin depolymerization reactions using a GO-UHP (Urea hydrogen peroxide) catalyst system.

358a Maleic Acid Hydrotropic Fractionation: Effective Deconstruction of Lignocellulose for Multi-Products Biorefinery

Junyong Zhu

Producing bioenergy, biochemicals, and biomaterials from renewable plant biomass can achieve a sustainable future through carbon sequestration and reduced use of fossil hydrocarbons. Developing lignocellulose fractionation processes that are not only sustainable, but also yield easily processable building blocks is the key to commercial success. For the past decade, we have devoted our efforts to advance the science and engineering critical to commercialization of biorefining technologies.

In this presentation, I will report our recent discovery of hydrotropic properties of a recyclable (low solubility) dicarboxylic acid, maleic acid (MA), for potential low-cost and efficient fractionation of lignocelluloses through rapid dissolution of lignin and hemicelluloses at atmospheric pressure. MA hydrotropic fractionation (MAHF) of wood results in two fractions: (i) a primarily cellulose rich water-insoluble solid (WIS) fraction that can be used for the production of high-value building blocks, such as lignocellulosic nanomaterials, fibers, and/or sugars through subsequent enzymatic hydrolysis; and (ii) a spent acid liquor stream containing hemicellulosic sugars and minimally condensed dissolved lignin that can be easily precipitated by simply diluting the spent acid liquor using water to below the minimal hydrotropic concentration. The dissolved hemicelluloses in the spent liquor can be dehydrated into furans using the MA in the spent liquor after lignin precipitation without additional catalysts. MA can then be reconcentrated, recovered, and reused. Nuclear magnetic resonance analyses revealed that MAHF dissolved lignin (MAHL) has a very high content of b-aryl ether linkages with minimal condensation, which facilitated catalytic conversion into monophenols. The MAHF is carboxylated with light color and therefore is suitable for other applications such as sunblock and skincare products. Carboxylation of the residual lignin in the fractionated WIS enhanced enzymatic saccharification by decreasing nonproductive cellulase binding to lignin, as well as facilitated nanofibrillation of WIS for producing lignin-containing cellulose nanofibrils (LCNFs) to substantially decrease fibrillation energy. MA is a FDA-approved indirect food additive (21CFR175-177) and can valorize three major lignocellulose components alone, it therefore has significant advantages for biorefinery applications.

358b Advances in Biomass Processing Solvents in Biorefinery Concept

Chang Geun Yoo

Environmental issues and shortage of fossil-based resources have increased the demand for alternative fuels, chemicals, and materials. Lignocellulosic biomass, a heterogeneous biopolymer mainly composed of polysaccharides (i.e., cellulose, hemicellulose) and a natural aromatic macromolecule (i.e., lignin), is a promising resource in these applications. Despite the enormous potential of lignocellulosic biomass as fossil fuel alternative, biomass conversion is still challenging due to its recalcitrance factors. Biomass pretreatment has been developed to facilitate and maximize the conversion of biomass. For improving the current conversion efficiency and expanding the utilization of biomass from a single component (i.e., cellulose) to entire biomass components in the biorefinery concept, pretreatment solvent plays a critical role. Lignin-derived deep eutectic solvent and bifunctional aromatic acid are recently developed to improve both the quality and quantity of polysaccharides and lignin in biomass by reducing the unwanted modification of hemicellulose and lignin during the processes. Technical features and obstacles of these pretreatment solvents will be presented with the remaining challenges and future perspectives.

254a Chemical Synthesis of Potential Prebiotic Oligosaccharides from Simple Sugars and Lignocellulosic Biomass in Concentrated Acids

Meijun Zeng, Ning Li, Theresa Astmann, Jee-Hwan Oh, Jan Peter van Pijkeren, Xuejun Pan

Prebiotics are non-digestible compounds that modulate the composition and/or activity of the gut microbiota through their selective fermentation by microorganisms in the gut, thus conferring health benefits to the host. Non-digestible oligosaccharides like fructo-oligosaccharides (FOS), galacto-oligosaccharides (GOS), isomalto-oligosaccharides (IMO), xylo-oligosaccharides (XOS), and manno-oligosaccharides (MOS) have been recognized as prebiotics, and they have been commercially produced for application in functional food/drinks and dietary supplements. It is highly attractive to develop efficient processes for large-scale production of prebiotic oligosaccharides from inexpensive and abundant substrates. For this purpose, we developed chemical methods for the synthesis of potential prebiotic oligosaccharides from biomass-derived simple sugars (bottom-up process) or directly from lignocellulosic biomass (top-down process) in concentrated acid systems. In the bottom-up synthesis strategy, gluco-oligosaccharides (GlcOS) were synthesized from glucose via dehydration condensation (acid-catalyzed glycosylation) in concentrated sulfuric acid (SA, 60–92%) with a high single-pass yield of GlcOS (83%) at 70 °C for 20 min. The synthesized GlcOS were a mixture of disaccharides (44%), trisaccharides (22%), tetrasaccharides (13%), and higher oligosaccharides (DP 5-9). The glucose residues in GlcOS were linked by predominantly α- and β-(1→6) linkages and other linkages including (1→4), (1→3), (1→2), and (1→1). The *in vitro* fermentation of GlcOS by probiotic bacteria *Bifidobacterium bifidum*, *Bifidobacterium animalis*, and *Limosilactobacillus reuteri* suggested that GlcOS could be utilized as a carbon source for bacterial growth, and their promotion effect was comparable to the commercial prebiotic IMO. It was found that concentrated phosphoric acid (PA, 70-85%) was as efficient as concentrated SA for GlcOS synthesis. Different oligosaccharides were also successfully synthesized from other mono- and di-saccharides (e.g., galactose, xylose, arabinose, lactose, maltose, and cellobiose) in the concentrated acid systems. In the top-down synthesis strategy, GlcOS and XOS/MOS were synthesized from cellulose and hemicelluloses (xylan or mannan) in lignocellulosic biomass (corn stover) via simultaneous hydrolysis and glycosylation in concentrated SA/PA.

254b Reaction Engineering *in Planta*? Tales of Mass Transfer Limitations and Their Kinetic Consequences at the Mesoscale

Nicholas E. Thornburg, Ryan M. Ness, Meagan Crowley, Lintao Bu, Michael Pecha, Francois Usseglio Viretta, Vivek Bharadwaj, Yudong Li, Xiaowen Chen, David A. Sievers, Edward Wolfrum, Michael G. Resch, Peter N. Ciesielski

Between the molecular and reactor scales—familiar favorites to the chemical engineering community—lies an intermediate regime termed the "mesoscale" where chemical reaction kinetics and transport phenomena compete along similar time and length scales. However, little is known about the coupled chemistry and physics of biomass conversion at the mesoscale, which govern effective rates of component extraction from plant cell walls during condensed-phase fractionation processes. First, we will introduce an experimentally validated simulation framework that determines transport-independent kinetic rate constants upon incorporating realistic feedstock characteristics for an exemplary hardwood solvolysis process. This generalizable mesoscale reaction–diffusion modeling approach will then be extended to validate and predict the alkaline deacetylation of corn stover, an emerging pretreatment system that removes acetyl moieties from hemicellulose prior to mechanical refining to improve downstream enzymatic sugar yields. Reaction–diffusion models are developed and validated for three major anatomical fractions (cobs, husks and stalks), and model findings categorize experimental feedstock performance into kinetic-controlled vs. diffusion-controlled regimes based on the particle size and microstructural attributes of each tissue type. Critically, we predict that typical corn stover particles as small as ~2.3 mm in length are entirely diffusion-limited for acetate extraction, with experimental effectiveness factors calculated to be 0.50 for such processes—size thresholds consistent with the unrelated hardwood solvolysis system first introduced. Researchers are therefore recommended to conduct kinetic evaluations of condensed-phase pretreatments using biomass particles smaller than ~0.2 mm in length to avoid feedstock-specific mass transfer limitations, particularly when a catalyst phase is also present. Overall, this presentation highlights opportunities to improve biomass fractionation and conversion via reaction engineering and provides actionable kinetic information to guide the design and scale-up of emerging biorefinery strategies.

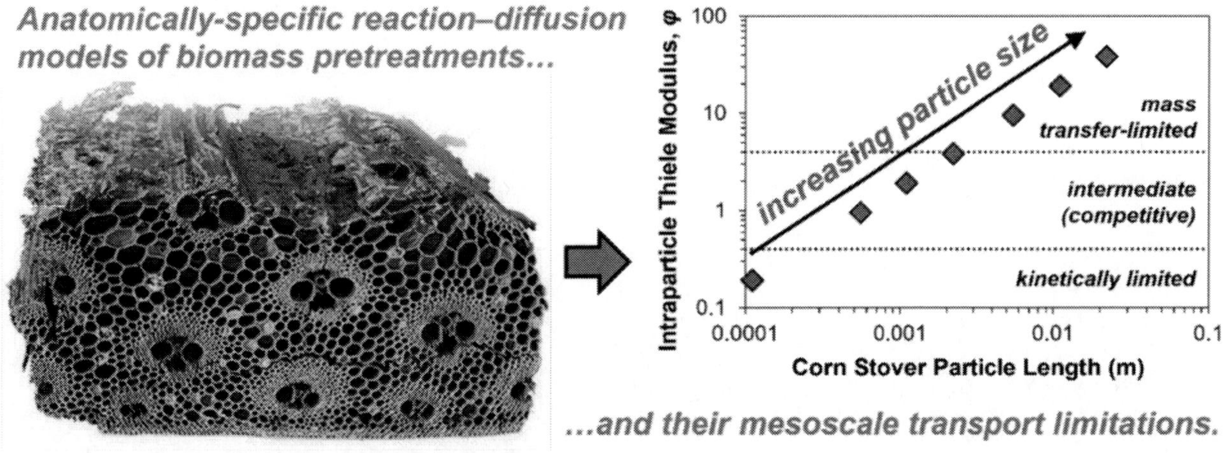

254c Catalytic Conversion of Bioresource to Graphene-Based Materials

Théotime Béguerie, Elsa Weiss-Hortala, Ange Nzihou

Graphene is a bidimensional material with one atomic layer as thickness. Graphene sheets are highly organized in graphitic structure or randomly oriented in turbostratic structure, leading to the formation of carbon nanotubes, carbon fibers, carbon black. Depending on the characteristics such as the length and the orientation of the graphene sheets, various properties could be developed: electrical conductivity, mechanical or thermal resistance. Therefore, graphene is considered as a high-performance material, in applications such as batteries, energy storage, electronics, and biology. Currently, materials are produced from petroleum-based industries (exfoliation or chemical vapor deposition) leading to a high negative environmental impact.

Graphene production from carbon resources requires the progressive organization of carbon atoms into fused benzene rings (namely basic structural units), then the coalescence for larger graphene sheets over 2000°C. Their organization (graphitic, turbostratic) will impact the final properties of the designed material. Aromatization and coalescence phenomena are strongly related to the original feedstock, and resources are sorted into graphitizable and non-graphitizable materials.

The aim is to use bioresources (biomass polymers), non-graphitizable carbons, to enhance the quantity and the quality of graphenic domains during catalytic pyrolysis up to 1800°C. For this purpose, the catalysts are more likely selected among inherent biomass minerals providing a green approach for graphene synthesis.

This study investigates the impact of catalyst on the evolution of the graphenic structure, texture and nanotexture in biochars. Graphitization experiments have been carried out in a high-temperature furnace (1800°C) with raw materials (lignin or cellulose) either impregnated or not with Ca or P catalysts. The remaining biochar was characterized to evaluate the quality of graphitization using combined techniques: HRTEM, Raman Spectroscopy and XRD. The impregnation of raw materials significantly improved the graphitization parameters such as length of basal sheets, stacking and significantly reduced the amount of amorphous carbon.

501a Microbial Upcycling of Post-Consumer Polyethylene Waste into Protein-Based Materials

Alexander Connor, Mattheos Koffas, Runye Zha

The overwhelming majority of plastics used today are derived from petrochemical feedstocks and are considered 'recalcitrant', meaning they are not readily degraded in most natural environments. Even with ideal application of conventional recycling strategies, these plastics can only be recycled a limited number of times before end-of-life disposal by incineration or dumping in landfills, leading ultimately to greenhouse gas emissions, pollution and disruption of natural ecosystems, and loss of feedstock material. Polyethylene is especially problematic, as it is predominantly used for short lifespan applications such as single use packaging. These challenges have led to a growing interest in processes that convert waste materials into higher value products that may share little to no chemical similarity with the starting material. This process, known as 'upcycling', has been widely applied to produce value-added fuels, chemicals, and materials, through methods such as pyrolysis and catalytic conversion reactions. Microbial upcycling is another strategy, which unlike the aforementioned approaches, has the ability to readily produce biodegradable or biologically active materials. Upcycling of plastics using microbial cultures has been used to produce several biodegradable materials, such as polyhydroxyalkanoate, fatty acids, and biosurfactants, but the field is lacking in documentation of upcycled protein-based materials. Within this context, silk proteins are some of the most desirable biomaterials to target for microbial upcycling due to their unmatched combination of properties that includes, but is not limited to, high toughness, biocompatibility, biodegradability, self-assembly capacity, thermal stability, solvent resistance, and ability to act as an optical waveguide. In addition, silk fibroins can be manufactured into a diverse array of constructs, such as coatings, fibers, and hydrogels, facilitating their use in a wide variety of applications. Developing an upcycling process to produce protein-based materials, such as recombinant silk, represents an innovation that can eliminate waste while simultaneously generating biodegradable value added materials.

Our work aimed to establish a microbial process for converting waste polyethylene into biodegradable recombinant silk proteins through use of the *Pseudomonas* genus. Several *Pseudomonas* species have demonstrated the ability to utilize the pyrolysis products of post-consumer polyethylene, which is chiefly comprised of alkanes, as a sole carbon source in flask cultures. Our work has uncovered factors that influence the growth of *Pseudomonas* when alkanes are used as the sole carbon source. These variables include the amount of alkane supplied, the amount and type of nitrogen source used, the addition of exogenous biosurfactant, and the presence of antibiotic selection marker. Results show cell densities as high as 7x109 cfus/ml can be achieved in a minimal media containing only hexadecane as a carbon source, which is comparable to cell densities achieved with common microbial media (such as Luria-broth). There is a substantial variation in the preferred culturing conditions among different *Pseudomonas* species, including that *P. aeruginosa* grows optimally in flasks with 1% w/v of alkane while optimal conditions for *P. oleovorans* include 0.1% w/v of alkane. Moreover, 1.9 g/L of ammonium nitrate is the preferred carbon source in this media for *P. aeruginosa* while 2.5 g/L of ammonium chloride is necessary for the optimal growth of *P. oleovorans*. Interestingly, the addition of exogenous biosurfactants to the cultures offers no benefit to the rate or overall amount of cell growth. Our work shows that the use of plasmid vectors can be used in a *Pseudomonas* host to achieve recombinant production of green fluorescent protein (GFP) and silk protein in Luria-broth. Plasmids have also been used to achieve the production of GFP in an alkane based media with a *P. oleovorans* host, at levels of 12 mg/L. However, plasmids are poorly maintained in the alkane based media, showing only 1-2% maintenance after 24 hours of culture time. Furthermore, the use of antibiotic selection severely inhibits growth in this media. Therefore, genomic integration of recombinant constructs was undertaken to improve the efficiency of the system and produce upcycled silk proteins, and represents a key strategy for future work in this space. Thus, this work expands the capability of microbial upcycling processes and contributes to future sustainability by demonstrating the first production of heterologous proteins from an alkane feedstock and reports on culturing parameters necessary to facilitate practical levels of cell growth and protein production when alkanes are supplied as a sole carbon source.

501b Development of Bioplastic: Need of the Hour

Rahul Taskar, Abhishek Yadav, Dr. Utkarsh Maheshwari

As a necessary component of our daily lives, plastics are becoming increasingly vital. The majority of commercially available plastics today are petroleum-based, which means that they can take hundreds of years or more to breakdown. As previously stated, degradation is not conceivable since the plastics generated are artificial polymers that are not recognised as food by the microorganisms that would be required to breakdown them.

Bioplastics, which are biodegradable polymers that contain components produced directly from renewable raw plant materials, have the potential to be a more ecologically friendly alternative to the currently non-degradable plastics used in consumer products. They are derived from plant materials and decompose into natural components that are inert in the soil and do not pose a threat to the environment. A number of bioplastics can degrade within a few weeks after being introduced.

The current research effort focuses on the creation of biodegradable plastics that are made from bio-based components, with the goal of making it more attractive to use biodegradable plastics in the first place. This breakthrough will herald a change in the way plastic is used since it will be based on an ecologically beneficial approach.

501c Renewable Barrier Polymers from Carbohydrate Nanomaterials: Processing and Properties

Yue Ji, Yang Lu, Meisha L. Shofner, Tequila Harris, J Carson Meredith

There is an ongoing international effort to develop circular alternatives to conventional plastics, particularly in the packaging sector. One way to achieve a circular economy of materials is to derive materials from renewable resources, waste materials, or recycled content. At the end of life, economical mechanisms should enable circulation of the carbon and other components back into production via chemical, physical or biological pathways. Plastic multilayer barrier primary packaging is a large contributor to unrecyclable or difficult-to-recycle waste. This talk describes innovations in production of barrier polymer coatings suitable for food, pharmaceutical and electronics packaging, based on combinations of cellulose- and chitin-based renewable nanomaterials. These abundant materials, sourced from plant biomass (forest or agricultural residue waste) and/or industrial food waste streams, are highly crystalline and have excellent potential as gas barrier materials for packaging. The talk will discuss four innovations in process-structure-property relationships: 1. mode of combination of CNCs and ChNFs: alternating layers versus direct blends, 2. mode of deposition: spray, blade and slot die coating, 3. modifications of films to manage water vapor barrier properties. Attention will be paid to the structure-property relationships and challenges to industrial implementation, including a discussion of future potential approaches to overcome these challenges.

501e Evaluating Sustainable Alternatives for Membrane Separation Processes: The Case of Poly Hydroxy Alkanoates (PHA)

Kseniya Papchenko, Eleonora Ricci, Maria Grazia De Angelis

Membrane technology is widely accepted to be a competitive alternative to traditional separation techniques, aiming to improve the overall sustainability of chemical processes. Still, most of the materials involved in the production of membranes cannot be considered sustainable, since they are fossil-based, and the possibilities to recycle them are somehow limited. This work aims to introduce a family of bio-based and biodegradable polymers as a sustainable alternative to materials traditionally used as membranes for gas separation.

Polyxydroxyalkanoates (PHAs) are a family of linear optically active semi-crystalline polyesters produced by bacterial fermentation, known for their fully renewable origin, biodegradability in many environments and physiological biocompatibility [1]. The gas transport properties of these materials are still scarcely characterized experimentally, and their determination is complicated by a number of uncertainty sources, such as a time-dependent degree of crystallinity. In this study, we aim at evaluating the structure-property relationships of different homo- and copolymers of the PHA family through molecular simulations, in order to gain information about their applicability in the membrane gas separation field. In particular, three homopolymers and two copolymers of the PHA family were considered:

- poly(3-hydroxybutyrate) (P3HB);
- poly(3-hydroxyvalerate) (P3HV);
- poly(4-hydroxybutyrate) (P4HB);
- poly(3-hydroxybutyrate-co-3-hydroxyvalerate) (PHBV);
- poly(3-hydroxybutyrate-co-4-hydroxybutyrate) (PHBB).

Molecular models of each material were simulated using Molecular Dynamics (MD), obtaining amorphous density and solubility parameter values that were successfully validated with experimental data found in literature [2, 3]. Other useful descriptors, such as accessible free volume and surface, were then evaluated and correlated to the chemical composition of different polymers.

The characterization of transport properties of the selected polymers was then initiated by evaluating sorption and diffusion values for two gases, CH_4 and CO_2, by means of Widom insertion method and MD simulations. The results were compared with experimental values, obtained through sorption tests, performed on PHBV with 8% of 3-hydroxyvalerate monomers, purchased from Merck-Sigma. The simulations showed how the transport properties in these materials are governed by the solubility-driven selectivity, throughout the whole composition range.

Furthermore, molecular simulations were used to obtain pressure-Volume-Temperature (pVT) data for the selected polymers that were fitted by different Equations of State. The parameters obtained from such a fitting were then used to extend the gas sorption predictions, giving a complete picture of transport properties for PHA family.

Molecular simulations provide a powerful screening method to identify the optimal biopolymer formulation for the desired gas separation process, e.g., CO_2/N_2, O_2/N_2 and CO_2/CH_4, allowing to minimize the experimental effort and to introduce new sustainable materials to the market.

References:

[1] K. Sudesh, H. Abe, Y. Doi, "Synthesis, structure and properties of polyhydroxyalkanoates: biological polyesters", Prog. Polym. Sci., 2000, 25, 1503-1555.

[2] H. Mitomo, N. Morishita, Y. Doi, "Structural changes of poly(3-hydroxybutyrate-co-3-hydroxyvalerate) fractionated with acetone-water solution", Polymer, 1995, 36, 2573-2578.

[3] H. Mitomo, W.-C. Hsieh, K. Nishiwaki, K, Kasuya, Y. Doi, "Poly(3-hydroxybutyrate-co-4-hydroxybutyrate) produced by Comamonas acidovorans", Polymer, 2001, 42, 3455-3461.

501f Synthesis of Thermoplastic Polyesters from Betulin, a Birch Bark-Derived Triterpenoid, Via Melt Polycondensations

Alexandra M. Lehman-Chong, Casey L. Cox, Joseph Stanzione III

Birch trees naturally exfoliate their bark in thin layers and strips that have historically been used for making multipurpose artifacts such as food containers and traditional medicine. However, on an industrial scale, birch bark is used as a low-grade fuel that is burned for energy production in the logging and papermaking industries. Birch bark contains natural chemical extractives, many of which have potential as bio-based molecules for chemical applications. Specifically, betulin is a pentacylic triterpenoid with a rigid, aliphatic chemical structure that can be extracted in appreciable quantities from birch bark. Because of its traditional medicinal uses, betulin is studied for its use in anti-inflammatory, anti-HIV, and anti-cancer pharmaceutical applications as well as a nutraceutical for products such as cosmetics. Due to its unique complex natural structure, betulin is also being studied as a bio-based monomer for preparing synthetic polymers. Here, betulin-based polyester thermoplastics were synthesized using a solventless melt polycondensation procedure. Betulin (bet) was used with bio-based linear aliphatic diacids azelaic acid (C9), 1,12-dodecanedioic acid (C12) and 1,18-octadecanedioic acid (C18) to synthesize thermoplastic polyesters poly(bet-C9), poly(bet-C12), and poly(bet-C18), respectively. Successful polyester syntheses were confirmed using ^{1}H-NMR spectroscopy. The synthesized thermoplastics exhibited M_w values ranging from 8,700 g mol^{-1} to 23,000 g mol^{-1}, measured using Advanced Polymer Chromatography. All polyesters were amorphous as per Differential Scanning Calorimetry with glass transition temperatures ranging from 78 °C to 130 °C and exhibited initial decomposition temperatures greater than 300 °C in nitrogen from Thermogravimetric Analysis. This work combines the use of betulin with plant-derived diacids to prepare polyesters with potentially 100% bio-based content. These investigations demonstrate that inherently nontoxic betulin can be used as a monomer to design thermoplastic polyesters that may pose less hazardous health effects upon degradation for multiple applications, including potentially for biomedical applications.

501g Biodegradable Composite Films from Carbohydrate Rich Food Waste Streams: Physicochemical Properties, Scale up Production, and Biodegradability

Shu Xu, Chaoyi Ba, Meltem Urgun-Demirtas

Plastic waste pollution in the environment has become a growing concern worldwide, which opened a new demand and market for biodegradable and/or compostable films to be used as single-use packaging across applications such as food and beverage, healthcare, personal care, and consumer goods. Developing suitable packaging materials from bioderived and biodegradable materials will create a more sustainable and cost-effective circular economy and alleviate the plastic accumulation problem in receiving environments. End-of-life testing of plastic materials in both natural (e.g., fresh water and soil) and engineered environments (e.g., composting and landfill facilities) is a great tool to estimate the fate of these plastics, hence identifying their persistence and the suitable pathways for their disposal.

This study aims to improve existing formulations and food waste extraction methods to synthesize and produce alternatives to single-use food packaging that are not degradable. We used banana and potato food waste extracts to synthesize new biodegradable blend films with chitosan, a film-forming polysaccharide extracted from crab and shrimp shells. These films showed good thermal stability up to 200 °C, elasticity with elongation up to 121%, and water vapor permeation properties comparable to chitosan films. These biofilms reached more than 50% biodegradation in freshwater, soil, and compost environments within 45–90 days and their estimated lifetimes in these environments range 90–260 days. We were able to scale up the production of these blend films to 10 L per batch using a bench top pilot-scale tape casting machine. The properties of blend films prepared at different production scales are compared. With excellent biodegradability in receiving environments and physicochemical properties needed for single-use applications, new films have great potential for single-use packaging applications.

501j The Integrated Production of Biopolymers, Monomers, and Electricity from Cheap Carbon Sources Using Microbial Fuel Cell Systems

Jianfei Wang, Shijie Liu

In recent years, microbial fuel cell systems have received extensive attention in the fields of bioenergy and environmental science. Different microbial fuel cell systems have been developed for power generation and wastewater treatment. At present, the application of microbial fuel cells is being extended to the integrated production of electricity and bioproducts. The anode and cathode chambers of microbial fuel cells can provide an ideal fermentation environment for microorganisms to produce valuable compounds. The sustainable production of biopolymers is also one of the current research hotspots. The biocompatibility and biodegradability of these polymer materials endow them with higher application value. Biopolymers are usually composed of one or more types of monomers, and these monomers are often used as important platform compounds in multiple industrial fields. Therefore, the production of monomers still needs to be further developed. This study focuses on the integrated production of monomers, polymers, and electricity using microbial fuel cell systems. The anaerobic environment provided by the anode chamber favors the lactic acid production by lactic acid-producing strains. The aeration in the cathodic chamber can provide an aerobic or microaerophilic environment for efficient accumulation of polyhydroxyalkanoates or their monomers, depending on the characteristics of certain strains. Correspondingly, the electron generation based on microbial metabolism in the anode chamber and the oxygen reduction in the cathode chamber also achieves electricity generation. During this process, the utilization of cheap carbon sources will also achieve the valorization of biomass resources.

501k Feedstock Design for High-Quality Biomaterial Manufacturing

Cheng Hu, Qiang Li, Arthur Ragauskas, Scott Sattler, William Rooney, Joshua Yuan

Traditional lignocellulosic feedstock design has focused on saccharification efficiency, which does not inform the material development. We have advanced fundamental understanding on how biomass structure characteristics impacts carbon material properties, which empowered the feedstock design toward high-quality materials. First, modeling of NIR data from over 2000 sorghum lines have identified 8 sorghum lines with diverse biomass structures. Carbon fibers were manufactured from these lines and underlying biopolymer characteristics defining crystalline structure and properties were identified. The study highlighted that lignin uniformity from the biomass defines the mechanical properties of carbon materials, which is very different from the traditional factors like lignin content and composition. Second, we further carried out cross-species studies with more diverse lignin structures from a broad range of biomass feedstock. Specifically, lignin extracted from hardwood (sugar maple), softwood (loblolly pine and red cedar), and herbaceous plant (corn stover and switchgrass) were used for carbon fiber manufacturing. Linear regression models were established to define the relationship among carbon fiber properties, microstructures, and properties. The results again highlighted that traditional biomass characteristics don't impact lignin carbon material properties. Instead, the content of β-O-4 linkage correlates significantly with tensile strength and elastic modulus of lignin carbon fiber, indicating that the more linear β-O-4 linkage would promote carbon fiber performance. We further demonstrated that a higher β-O-4 content also promotes electroconductive performance of carbon fiber and improves the crystalline content and size, which could be due to the better miscibility with the guest polymer. Thirdly, based on these fundamental understanding, we have engineered lignin biosynthesis in sorghum to achieve improvement of lignin molecular weight and linearity, which increased carbon fiber properties significantly. The fundamental understanding has paved path to use synthetic biology to design feedstock for various high-quality carbon and polymer materials. The value-added products will in turn empower cost-effective and sustainable lignocellulosic biorefinery.

501i Synthesis of Lignin-Based Polyesters

Hoyong Chung, Sundol Kim

Lignin is an abundant plant-based biopolymer that possesses high concentration of aromaticity which enhances mechanical and thermal properties. The sustainable raw material lignin is produced from a paper industry, where lignin is disposed of as an unwanted byproduct. In addition to its abundance and inexpensive supply, lignin is favorable for its numerous attractive properties, such as biodegradability, antioxidant activity, high carbon content, high thermal stability, and stiffness. These important features of lignin can be synergistically combined with the advanced functionalities of other polymers via covalent bond linkages. This presentation will report poly(lignin-co-poly(ethylene brassylate)) (poly(lignin-co-PEB)). PEB is non-human food vegetable oil, castor oil, based polymer. The poly(lignin-co-PEB) was prepared by two steps: 1) lignin modification to covert hydroxyl group and 2) covalent linking of lignin to PEB. However, there was a serious concern about the linearity of the proposed lignin-based polymers due to lignin's multiple hydroxyl reaction sites that lead uncontrolled crosslinking. The highly crosslinked polymer does not show thermoplasticity, and therefore is disadvantageous for scalable thermal processing. The talk will discuss solutions of this concern. During the initial lignin modification, the unwanted hydroxyl groups on lignin can be deactivated by capping agents. Then, only very small numbers of hydroxyl group remain in the lignin to avoid unwanted crosslinking. The resulting copolymer, poly(lignin-co-PEB) shows highly controllable thermal properties which is a thermoplastic behavior. In conclusion, the number of carboxylic acids on the modified lignin is precisely controlled to yield a lightly crosslinked (or linear) polymer that is a thermoplastic.

452a Using Artificial Neural Networks to Estimate Xylose Conversion and Furfural Yields in Autocatalytic, Organic/Aqueous Solvent Systems

Adam L. Job, Sarah Stratton, Charles E. Umhey, Karlene A. Hoo, Stephanie Wettstein

Artificial neural networks are becoming more commonly used in order to estimate outputs based on important inputs. In this work, an ANN was built to predict xylose conversion and furfural yield in autocatalytic organic/aqueous reactions that explained over 90% of variance for each output. This ANN allows for methodical solvent selection versus the traditional "guess and check" method, which reduces solvent waste and is better for the environment.

Platform chemicals such as furfural can be derived from biomass and present a valuable source as intermediates for fuels and chemical commodities. The dehydration reactions that form furfural can be catalyzed in a hydrothermal system as the result of hydronium ions from high temperature water and reaction byproducts, such as formic acid. In this work, we created a feed-forward artificial neural network (ANN) that estimates the xylose conversion and furfural yield from the reaction severity, which is a function of temperature and time, as well as the Hansen Solubility Parameter (HSP) for polarity for reaction systems consisting of organic/aqueous solvent mixtures. Using both severity and polarity as independent input variables, the percent variance explained by the ANN model was 95.5% for xylose conversion and 92.1% for furfural yield. In the case of xylose conversion, only one predicted point was outside of the 95% prediction interval and for furfural yield, only two points were outside. Importantly, the 10 experimental points that were not used to train the ANN were all estimated within the 95% interval, which indicates the ANN model captured the nonlinear correlations between the input and output variables for the entire experimental range of polarity and severity.

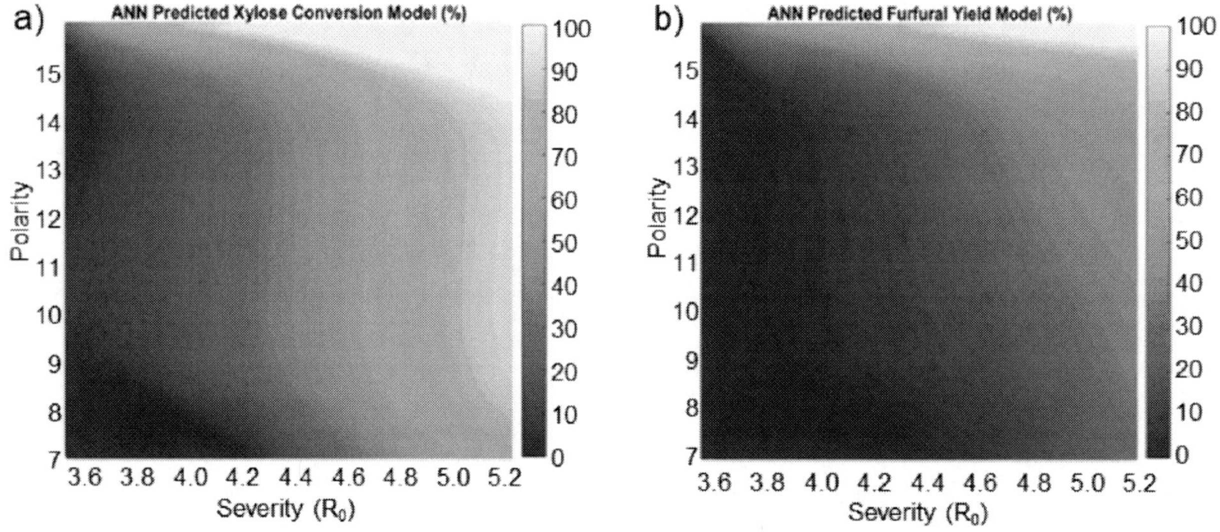

452b Valorization of Bioenergy Crops for Recovery of Anthocyanins and Sugars

Shivali Banerjee, Erik Sacks, Vijay Singh

The increased awareness for eco-friendliness and sustainability has shifted the interest of stakeholders from synthetic colorants to natural plant-based pigments. In order to meet the large-scale demand for natural pigments in industrial applications, the process should be economically viable and environmentally sustainable along with abundant supply of the feedstock. The presence of industrially relevant pigments (mainly anthocyanins) in bioenergy crops such as miscanthus, sorghum and sugarcane[1-5] has been reported in literature. The red pigment present in the leaf sheaths of *sorghum*, is rich in 3-deoxyanthocyanins, which are known for their stability to heat treatments and pH changes, making them a valuable source of natural colorants for industrial applications. In the case of sugarcane and miscanthus, the color is mostly concentrated in the stem and could be extracted before processing the crop for production of biofuels. In this study, miscanthus was analyzed for its total anthocyanin content (TAC) and structural carbohydrate composition. TAC was determined by a two-staged extraction method. The dried and ground biomass sample was mixed with acidified ethanol and was continuously stirred at room temperature for 1 h. The mixture was filtered and the filtrate was analyzed for the anthocyanin content by the pH differential method[6]. Solids retained on filter paper were recovered and extracted using acidified ethanol for 2 h for determining remaining anthocyanin content. The TAC was obtained as the sum of the anthocyanin content from the two sequential extractions and was in the range of 10.2 to 73.4 μg/g of biomass. The concentration of glucan and xylan in miscanthus ranged between 37.6 to 43.7% and 18.2 to 26.4%, respectively. After the extraction of anthocyanin, the residual biomass could be enzymatically hydrolyzed into sugar monomers which could be fermented to biofuels. Recovering anthocyanins from bioenergy crops in an integrated biorefinery would increase the overall profitability due to the increased revenue from this high-value co-product.

References

1. Integrated Biorefinery for Valorization of Engineered Bioenergy Crops—A Review. *Industrial Biotechnology* **2021,** *17* (5), 271-282.
2. Zhao, Z.; Yan, H.; Zheng, R.; Khan, M. S.; Fu, X.; Tao, Z.; Zhang, Z., Anthocyanins characterization and antioxidant activities of sugarcane (Saccharum officinarum L.) rind extracts. *Industrial Crops and Products* **2018,** *113*, 38-45.
3. Kayodé, A. P. P.; Nout, M. J. R.; Linnemann, A. R.; Hounhouigan, J. D.; Berghofer, E.; Siebenhandl-Ehn, S., Uncommonly High Levels of 3-Deoxyanthocyanidins and Antioxidant Capacity in the Leaf Sheaths of Dye Sorghum. *Journal of Agricultural and Food Chemistry* **2011,** *59* (4), 1178-1184.
4. Petti, C.; Kushwaha, R.; Tateno, M.; Harman-Ware, A. E.; Crocker, M.; Awika, J.; DeBolt, S., Mutagenesis Breeding for Increased 3-Deoxyanthocyanidin Accumulation in Leaves of Sorghum bicolor (L.) Moench: A Source of Natural Food Pigment. *Journal of Agricultural and Food Chemistry* **2014,** *62* (6), 1227-1232.
5. Li, X.; Yao, S.; Tu, B.; Li, X.; Jia, C.; Song, H., Determination and comparison of flavonoids and anthocyanins in Chinese sugarcane tips, stems, roots and leaves. *Journal of Separation Science* **2010,** *33* (9), 1216-1223.
6. Lee, J.; Durst, R. W.; Wrolstad, R. E., Determination of Total Monomeric Anthocyanin Pigment Content of Fruit Juices, Beverages, Natural Colorants, and Wines by the pH Differential Method: Collaborative Study. *Journal of AOAC INTERNATIONAL* **2005,** *88* (5), 1269–1278.

452c Effect of Fines % on Ultrasonic Dewatering of Cellulose Nanofibrils

Udita Ringania, Robert Moon, M. Saad Bhamla

During the production of cellulose nanofibrils, mechanical fibrillation leads to the presence of different % of fines in the final product, defined as the fibrils under 200 um. Although the presence of fines has demonstrated mechanical advantages in composite materials, it also increases the water retention by the fibrils leading to increased dewatering time and energy. As addressing the high energy requirements during dewatering and drying of CNF is a high-priority target in the community, understanding the effect of underlying fibril dimensions on water retention becomes essential. In the present talk, we will discuss the effect of fines % of CNF on the dewatering efficiency of the ultrasonic dewatering technology and compare this with traditional dewatering/drying technique. Ultrasonic dewatering technique leverages the instabilities at the air-liquid interface generated by high frequency sound waves leading to the subsequent removal of water from the CNF slurry as cold mist. We will also discuss the redispersion of the subsequent dewatered samples and present an optical microscopy characterization technique to differentiate amongst various fines% in a fast and easy way. This microscopic analysis is also supported by nanoscale characterization through SEM.

452d Characterization of Brewer's Spent Grain through Thermogravimetric Analysis for Energy Production

Spencer Fogelquist, David Wagner

Introduction

Brewer's spent grain (BSG) makes up 85% of the brewing industries waste [1]. The waste product can be repurposed as a biofuel for energy production. The energy produced would be more sustainable than other fuel sources such as coal. This is due to using the waste product as a fuel source compared to decomposition in a landfill where methane is released into the atmosphere. Being sustainable is beneficial to breweries because they can increase profits with new product to sell and being able to get tax credits for sustainable practices.

Characterization of BSG will provide information in how the BSG devolatilizes and what the activation energies are based on kinetic behavior. The characterization will focus on char yield from the vary thermal conditions being studied. Thermogravimetric analysis (TGA) will be used to study the reaction rate of devolatilization and determine the activation energy [2], [3]. The activation energy from the TGA will be compared with Flynn-Wall-Ozawa method. Differential scanning calorimetry is used to analyze how char properties change during devolatilization [4].

Method/Experimental (may need rewording, same as other proposal)

Two-row brewers malt was obtained from Briess Malt andamp; Ingredients Company. The grain was milled via a 2-pin roller mill. Once the grain was milled, 1.25 quarts of water was heated to 77 °C in a 2-quart pot. Once the water was at 77 °C, one pound of the milled 2-row brewers malt was added slowly to the water. Due to thermal differences this brought the waters temperature down to 69 °C. The pot of water and grain was then kept at 69 °C ± 5 °C for 1 hour. Following the hour of heating, the grains were drained of water and were placed in a 105 °C oven for 24 hours. After the grains were cooled down, they were milled twice via the 2-pin roller to break apart the agglomerations of dried grain and to create larger a distribution of the particle sizes. The grain particles were then sieved and separated into ranges of 500-600 μm, 250-300 μm, and 125-150 μm.

Thermogravimetric Analysis

The sample procedure had four steps applied to each run. First, the TGA system is equilibrated at 40 °C. Next, a gas feed mixture of carbon dioxide and air flow into the TGA. This step is followed by a linear ramp step with varying heating rates and varying final temperatures. The last step is to hold the sample at the final temperature for 30 minutes to observe isothermal effects at final temperatures. The TGA was manufactured by TA Instruments, model 5500. Table 1 shows the design of experiments (DoE) approach used to characterize the potential of spent brewer's grain for use in energy production applications. Measurements (responses in the DoE) include char and volatiles yields, maximum derivate weight and its corresponding temperature, and the rate at 25 and 50 % mass loss for activation energy and reactivity determinations.

Results and Discussion

Figure 1 shows the direct effects of the different factors on how char yield responded in non-isothermal heating conditions. As the final temperature increased, the char yield decreased [5]. As heating rate increased, the char yield increased [6]. As particle size increased, the char yield increased [7]. As the

CO_2 volume percentage in air increased, the char yield increased as well [8]. The major factors controlling char yield are the final temperature and the heating rate used to devolatilize BSG. Particle size was another major factor, but was a non participant if the final temperature was high enough and heating rate was slow enough to allow full devolatilization. CO_2 was a minor factor as it only slightly increases char yield with higher CO_2 percentages.

Figure 2 shows that direct effects of the different factors on char yield for isothermal runs. As heating rate increased, char yield decreased. As the isothermal temperature increased, the char yield increased. As the particle size increased, there was a slight decrease in char yield. As the CO_2 percent in air increased, char yield increased as well. The major factors relating to char yield were the heating rate and isothermal temperature. Particle size and CO_2 percentage in air were both minor factors in char yield.

Comparing the two different environments in the modulated TGA runs, the activation energies between 125-150 μm and 250-300 μm, had no difference between how they reacted in the CO_2 rich environment compared to the pure air system. There was a slight decrease in activation energy for the 500-600 μm particle sizes when in CO_2 rich environment compared to the air environment. The slight decrease in activation energy can be seen as negligible change as it was the largest particle size where mass transfer can cause differences in devolatilization. CO_2 does not show major effects on the activation energy of material during devolatilization. This can be observed in Table 2.

Conclusion

Char yield between non-isothermal and isothermal BSG devolatilization had inverse relationships. Non-isothermal systems showed char yield increased when heating rate, particle size increased, and CO_2 percentage in air increased. Char yield decreased when the final temperature was increased. Isothermal systems showed that char yields increased with final temperature and CO_2 percentage in air increased. Char yield decreased when heating rate increased and particle size decreased. The factors in order of significance were final temperature, heating rate, particle size, and CO_2 percentage in air according to JMP statistical analysis. Modulated TGA showed no difference between activation energies when comparing air and CO_2 rich environments.

References

[1] A. S. N. Mahmood, J. G. Brammer, A. Hornung, A. Steele, and S. Poulston, "The intermediate pyrolysis and catalytic steam reforming of Brewers spent grain," *J. Anal. Appl. Pyrolysis*, vol. 103, pp. 328–342, 2013, doi: 10.1016/j.jaap.2012.09.009.

[2] A. O. Balogun, F. Sotoudehniakarani, and A. G. McDonald, "Thermo-kinetic, spectroscopic study of brewer's spent grains and characterisation of their pyrolysis products," *J. Anal. Appl. Pyrolysis*, vol. 127, no. April, pp. 8–16, 2017, doi: 10.1016/j.jaap.2017.09.009.

[3] P. Li, L. Teng, M. Guo, M. Zhang, and S. Seetharaman, "Oxidation kinetics of AlN under CO 2 atmosphere," *Metall. Mater. Trans. B Process Metall. Mater. Process. Sci.*, vol. 43, no. 2, pp. 406–412, 2012, doi: 10.1007/s11663-011-9611-5.

[4] S. Bibi *et al.*, "A preliminary investigation to group disparate batches of licit and illicit diazepam tablets using differential scanning calorimetry," *Anal. Methods*, vol. 7, no. 20, pp. 8597–8604, 2015, doi: 10.1039/c5ay01711d.

[5] Y. Zhang *et al.*, "Characteristics of biomass fast pyrolysis in a wire-mesh reactor," *Fuel*, vol. 200, pp. 225–235, 2017, doi: 10.1016/j.fuel.2017.03.070.

[6] R. sheng Xu, J. liang Zhang, G. wei Wang, H. bin Zuo, P. cheng Zhang, and J. gang Shao, "Gasification behaviors and kinetic study on biomass chars in CO2 condition," *Chem. Eng. Res. Des.*, vol. 107, pp. 34–42, 2016, doi: 10.1016/j.cherd.2015.10.014.

[7] A. Zolghadr, J. J. Biernacki, and R. J. Moore, "Biomass Fast Pyrolysis Using a Novel Microparticle Microreactor Approach: Effect of Particles Size, Biomass Type, and Temperature," *Energy and Fuels*, vol. 33, no. 2, pp. 1146–1156, 2019, doi: 10.1021/acs.energyfuels.8b03395.

[8] N. S. Yüzbaşı, "Pyrolysis and Combustion Behaviour of Various Fuels in Oxygen-Enriched Air and CO2 Atmospheres," no. February, p. 122, 2011.

Table 1: Design of experiments factors and levels for characterization of brewer's grain biomass.

Factors	Low	Medium	High
Heating Rate (°C/min)	5	15	45
Particle Size (um)	125-150	250-300	500-600
Non-isothermal Temperature (°C)	500	600	700
Isothermal Hold Time (min)	0	-	60
CO_2 % in air	10	20	30

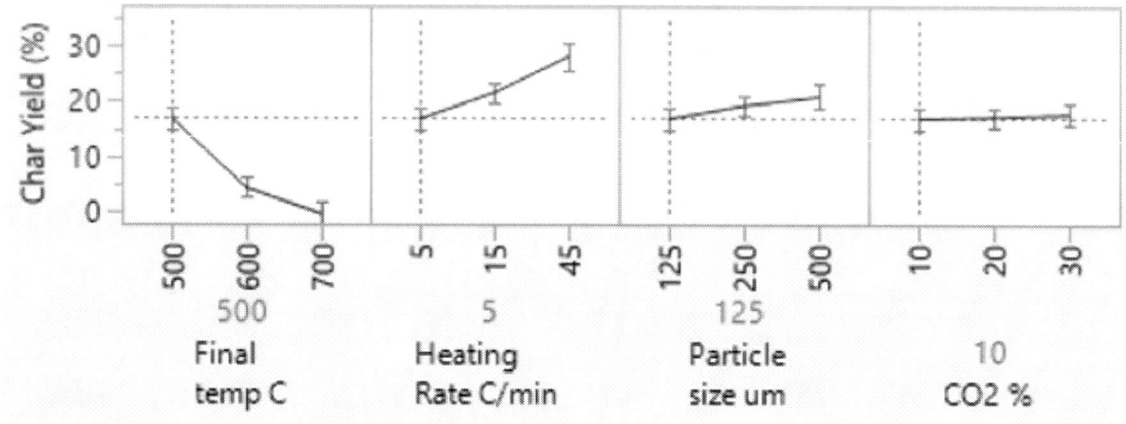

Figure 1. Non-isothermal parameter effects on char yield

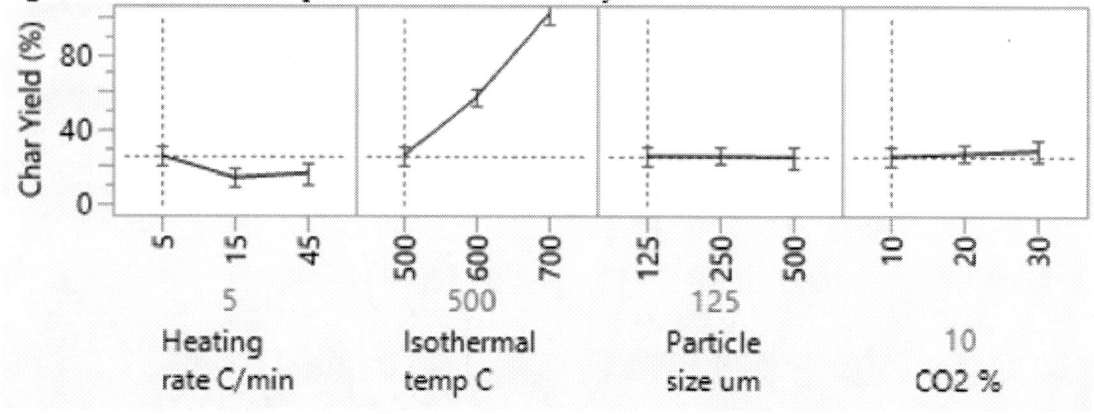

Figure 2. Isothermal paramter effects on char yield

Table 2. Modulated TGA Activation Energy comparing varying particles sizes with air and 20% CO_2 in air environments.

Particle Size (μm)	Activation Energy (kJ/mol) at 50% mass loss in air environment	Activation Energy (kJ/mol) at 50% mass loss in CO_2 rich environment
125-150	174	174
250-300	174	174
500-600	184	179

452e High-Grade Lignin Production: Evolution of Lignin Structure through γ-Valerolactone-Assisted Hydrolysis of Biomass

Feng Cheng, Sarah Liu, Steven Karlen, Hoon Kim, Fachuang Lu, John Ralph, George Huber, James A. Dumesic, Leoncio Santiago Martínez

High-grade lignin, as a renewable material, shows potential value in the products market (e.g., vanillin, bio-based polymers, and carbon fiber). However, the development of high-grade lignin is currently limited by the high cost and small scale of production. γ-Valerolactone (GVL) is a renewable and effective cosolvent to assist biomass hydrolysis and lignin extraction from biomass. Tuning the operating conditions (80–120°C and 1–24 h) can increase the yield of lignin but can also deteriorate its native structure. Our work aims to improve the yield as well as quality of lignin by developing an effective strategy to reach high yields while simultaneously maintaining a high grade of lignin. A variety of techniques, including nuclear magnetic resonance (NMR) spectroscopies (solution- and gel-state ^{1}H–^{13}C 2D HSQC and ^{31}P) and gel–permeation chromatography, were applied to characterize the structures of (1) the lignin in virgin biomass, (2) the lignin extracted into the GVL phase, and (3) the lignin remaining in the cellulose-rich solid phase. The crucial β-ether content, the aliphatic hydroxyl group content, and the molecular weight of various lignin fractions indicate that the native structure of lignin remaining in the cellulose-rich solid phase was more similar to native lignin compared to lignin dissolved in the GVL phase under the same conditions. As guided by the above observation, a comparison was carried out between a single-step 90 °C-3 h process and a two-step 90 °C-1.5 h treatment. The results show that the two-step process gave slightly higher lignin yield (31% vs. 29%) with a higher β-ether content (51% vs. 44%). The results demonstrate that multiple cycling of biomass for cosolvent-assisted hydrolysis can improve both yield and quality of lignin. Most importantly, such a strategy can be realized in a large-scale continuous reaction system, which will be conducive to commercialization of high-grade lignin production.

452f Enhanced Feedstock Characterization and Modeling to Facilitate Optimal Preprocessing and Deconstruction of Corn Stover

Dylan Cousins, William Otto, Asif Hasan Rony, John E. Aston, David Hodge

Chemical and physical heterogeneity in herbaceous biomass feedstocks such as corn stover are due to substantial differences in plant tissue types. These differences can contribute significant challenges to handling, preprocessing, and conversion in biorefining processes. A second, related challenge is the problem of quantifying this heterogeneity within feedstocks as it relates to processing in a biorefinery (*e.g.*, handling in screw feeders, response to particle size reduction, response to pretreatment and enzymatic hydrolysis, solid-liquid separations). In this talk, we will describe results on novel biomass fractionation technologies and developing improved characterization tools for assessing biomass heterogeneity and the responses to fractionation. Different corn stover anatomical fractions (cob, rind, stem pith, leaves, *etc.*) are shown to exhibit significant differences in chemical composition, bulk physical properties and respond differently to pretreatment and enzymatic hydrolysis. Through combinations of comminution, sieving, and air classification, we demonstrate several preprocessing strategies for recovering corn stover fractions enriched or depleted in different anatomical components that have the potential to facilitate streamlined processing in a biorefinery. Next, we develop physics-based predictive models for predicting the partitioning of corn stover anatomical components during air classification by differences in properties and validate that these models can be used to optimize these separations to achieve target processing outcomes. Finally, by coupling near-infrared (NIR) spectroscopy to machine learning tools, we develop models that can be used to predict not only composition, but the relative abundance of anatomical components.

634b Simultaneous Heavy Metal Ion Capture and Detection Via Bio-Based Triboelectric Sensor

Geng-Sheng Lin, Haiyang Zou, Lan Gan, Zhaohui Tong, Yong Ding, Yongsheng Chen

With the rapid development of technology, various pollutants are created and discharged into the environment. Among them, heavy metal wastewater is carcinogenic and lethal, which poses a significant threat to the environment and public health. However, current prevailing methods for detecting heavy metals require expensive equipment and arduous operation. Therefore, a green and efficient method is urgently needed to detect and remove heavy metal ions simultaneously from wastewater. Herein a cellulose-based membrane was fabricated and can effectively detect and capture heavy metal ions simultaneously. After grafting different functional groups, membranes can selectively capture heavy metal ions in the solution. Furthermore, with the incorporation of a triboelectric device, the device can then differentiate heavy metal solutions with different concentrations. This study also shed light on the adsorption and triboelectric mechanisms. To sum up, this bio-based membrane shows excellent potential in practical heavy metal wastewater treatment and paves future advancements in heavy metal detection systems.

634c Conversion of Lignocellulosic Biomass to Highly Functional Micro- and Nano-Materials for Water Treatment

Mica Pitcher, Amir Sheikhi, Breanna Huntington, Juliana Dominick

An important pillar of sustainable development is the preparation of functional materials derived from renewable resources. Nevertheless, facile methods to convert lignocellulosic biomass into value-added, highly functional micro- and nanomaterials remain limited. Here, we evaluate the potential of a sustainable method to convert a variety of lignocellulosic biomass into highly charged products. We show how the sequential oxidation of both delignified (e.g., softwood pulp and cotton) and untreated (e.g., corncob and tomato peel) lignocellulosic sources yield three distinct products, including nanoparticles, fully solubilized polymers, and microproducts, bearing up to 6 mmol of carboxylate groups per gram, which is ~500% beyond the theoretical charge content of cellulose nanocrystals (CNC). The main difference among the products was the type of nanoparticles: delignified sources yielded hairy cellulose nanocrystals, whereas lignin nanoparticles were formed from the untreated sources. As a proof-of-concept for the applications of these materials, we use carboxylated cotton microproducts to remove ~ 92.5 mg of neodymium (Nd) per gram from a dilute Nd solution in less than 5 min and recover approximately 64% of it via pH adjustment. This work may provide new opportunities for the conversion of a wide array of lignocellulosic biomass into highly functional biocolloids with the highest charge density ever reported for advanced sustainable applications.

634d Tune Cellulose Nanocrystal Alignment By Combining the Additions of Electrolytes with Shear-Based Alignment

Ananya Ghosh, ZhongYang Cheng, Zhihua Jiang

Cellulose nanocrystal (CNC) is a renewable, non-toxic and biodegradable material. It is produced from inexpensive renewable biomass and exhibits several unique chemical, optical, electrical, and electromechanical properties. To fully utilize these unique properties, CNCs have to be aligned with a high degree. However, the self-assembled CNC film exhibits low transparency due to its chiral nematic structure. When a simple mechanical shearing is applied, CNCs can be aligned so that CNC films are transparent with an enhanced chemical barrier property. It is also expected that CNC film with a high orientation degree exhibits a piezoelectric effect. A higher shear rate can result in a higher degree of alignment. However, CNC film made using a higher shear rate include more cracking and localized defect. Additionally, very higher alignment is not easy to achieve by a single mechanical shear stroke when dealing with CNC gel.

The aim of this project is to develop a CNC-based material from CNC gel which can be used as a coating or piezoelectric material without the limitations mentioned above. With the addition of electrolytes, the alignment degree of CNC can be significantly enhanced. For example, for a shear rate of 1,000 s^{-1}, the alignment degree increases from 32% without electrolyte to 90% with electrolyte. In addition, it is observed that adding electrolyte also significantly improve the flexibility of CNC films. This study also provides insights into the effect of the experimental conditions (i.e., the type of electrolytes, shear rate, and thickness) on the mechanical, optical, rheological, barrier, and piezoelectric properties of CNC films.

634e Predicting the Morphological Properties of Activated Carbons Produced from Lignocellulosic Materials

Daniel Meadows, Delaney Clouse, Sushil Adhikari, Virginia Davis

Every year approximately 200 billion tons of biomass waste is produced. Identifying value added products that can be produced from waste biomass is important for both economic viability and sustainability metrics for the forest products, agricultural, and other biomass industries. Numerous researchers have explored producing activated carbon from various biomass sources ranging from palm fronds to leftover food. However, few studies have explored how key constituents of lignocellulosic biomass affect the properties of activated carbon. In this work, an augmented second-order simplex lattice mixture design was used to explore how the relative concentrations of cellulose nanofibers (CNFs), cellulose nanocrystals (CNCs), and alkali lignin (LIG) affected the properties of activated carbons produced from a two stage carbonization and activation process. The response variables included yield, specific surface area and micropore fraction. The specific surface area and micropore fraction were determined by BET analysis and the *t-plot* method. The mixture regression models revealed clear relationships between all three activated carbon characteristics and biomass precursor composition. The highest yield of 43% was obtained from pure lignin, the most thermally stable material. Pure CNF resulted in both the highest micropore fraction, ~80%, and the lowest specific surface area, 350 m^2/g. The highest observed specific surface area of 1940 m^2/g was obtained from a precursor composition of 17/17/66 % (by mass) CNF/CNC/LIG blend. The models resulting from the experimental design were validated by producing activated carbon from a 15/35/50 % (by mass) CNF/CNC/LIG blend. This composition was chosen by equally weighting the desirability of maximizing both surface area and micropore fraction. The models predicted that the activated carbon would have a specific surface area of 1320 m^2/g and a micropore fraction of 60%. The experimentally measured values were 1240 ± 120 m^2/g and 58 ± 1%. The percent differences amongst the predicted and experimental values were 6% and 3% for surface area and micropore fraction respectively. This result highlights that the mixture design was effective in enabling the development of a robust model that can be used to design activated carbon precursor mixtures to maximize the most desirable properties for a given application. Moreover, understanding the role of individual biomass components in activated carbon precursors will aide in determining what types of biomass waste will be most effective for producing activated carbons for applications requiring different properties such as adsorbents for environmental contaminants and electrochemical devices.

667a The Drive for Greener Materials and Sustainability Leadership

Debbie F Mielewski

Since the first laboratory success and implementation of soy-based foam 15 years ago, the Ford sustainable materials research team has been expanding the use of industry-first, sustainable plastic materials in vehicles. Additional successes have been around natural fiber-reinforced composites such as wheat straw, rice hulls, and cellulose fiber from trees, which have been found to meet all of the stringent requirements of automotive components while significantly reducing weight and environmental impact. Many times, we've discovered that natural materials exceed the performance of traditional plastic composite materials, having higher heat, vibration, or sound-absorbing performance. Realizing the importance of establishing a circular economy, we've intensified our search for industrial partnerships to utilize waste streams in value-added automotive applications. In 2019, Ford with partner McDonald's was the first to launch coffee chaff reinforced plastic on the headlamp housings of the Lincoln Continental. The material saved 1 pound per vehicle and had a 20°C higher heat performance, a perfect material enhancement match for heat-generating bulbs. And during the pandemic, Ford and partner HP launched an industry-first injection molded part made from 3d printing waste powder. Last year, Ford introduced parts made from 100% ocean fishing nets on the Ford Bronco as a response to address ocean plastic accumulation. During her talk, Dr. Mielewski will discuss the broad portfolio of greener materials available now and those anticipated for the future that can not only reduce the automotive industry's footprint but many other industries as well.

667b Tree Bark-Derived Lignin Containing Nanocellulose Fibrils for Enhancing Performance of Plastic Composites

Ning Yan

In recent years, we have developed novel lignin containing nanocellulose fibrils (LNCF) from tree bark residues in sawmills. LCNFs showed great promises as renewable lignocellulosic nanobiofillers for making high performance plastic composites. With the presence of lignin on the surface, LNCFs exhibited some distinctive characteristics compared to conventional nanocellulose fibrils without lignin. The processing method for producing LNCFs using bark residues was simple and required low energy input. We have demonstrated excellent potentials of applying bark derived LNCFs for reinforcing engineering plastics as well as making fully biodegradable plastic biocomposites with highly improved mechanical properties and functional performances. Melt blending of spray dried LNCFs with polypropylene showed the resulting composites having significantly improved water-repellence, thermal stability, and tensile properties compared to composites made with nanocellulose fibrils without lignin. With LNFCs additions to PLA matrix, higher mechanical, thermal, and water vapour barrier properties were observed for the fully biodegradable biocomposites due to effective interfacial coupling. Our study has showcased that LNCFs derived from forestry residues can act as highly effective reinforcing nanofillers well-suited for making high performance sustainable plastic and biodegradable composites.

667c Fabrication of Lignocellulose-Based Materials for Environmental, Energy, and Packaging Applications

Xuejun Pan

This presentation is to report our recent studies on the fabrication of lignocellulose-based materials (nanocrystals, hydro/aerogels, and films) using a procedure of dissolution/dispersion and regeneration and their applications in environmental, energy, and packaging areas. Because of the capability of swelling and dissolving cellulose, molten salt hydrate (MSH), such as lithium bromide trihydrate (LBTH, ~61 wt% aqueous solution of LiBr), was used as a solvent to fabricate the materials. The dissolution of cellulose in LBTH is dependent on LiBr concentration, temperature, pH, and cellulose properties (consistency, particle size, and degree of polymerization). It is the ion-dipole coordination between hydrated Li^+ and cellulose hydroxyl that disrupts the intermolecular H-bond of cellulose and then results in the dissolution of cellulose in the LBTH. The swollen and/or dissolved cellulose in the LBTH is regenerated when cooled down and washed with water, in which the LBTH can be recovered for reuse. The cellulose polymorph is transformed from I to II during the processing. The fabricated materials can be used for the metal and formaldehyde removal from water and indoor air, the energy harvesting by triboelectric nanogenerator (TENG), and food packaging, respectively.

667d Value Addition of Waste Plastics through Upcycling for a Sustainable Future

Amar K. Mohanty

At present ~450 million tons of plastic per year is produced globally which is expected to exceed 1 billion tons by 2050. Additionally, each year globally 300 million tons of plastic waste is generated which is equivalent to the weight of the entire human population (UN report). It is estimated that 9 billion tons of plastic waste will accumulate by 2050 which equates to around one ton of plastic waste being created by each person around the globe. Plastic waste is prevalent as it is omnipresent in the atmosphere, land, oceans and even in the human blood. Make-use-through type plastic economy (linear model) is not sustainable; thus, we must opt for a circular economy approach. The recyclability or circularity of plastic waste can be targeted and altered to have an overall economic benefit for society and the environment. The mixed nature of waste plastics and their complex compositions limits their recyclability, with more than 50% not being recyclable. Further, the plastic waste that is recyclable results in less value, being counterproductive. In recycling, the final product targets at least the same or original value. Down-cycling results in lower value whereas in upcycling the final product can generate higher value. Upcycling helps in waste valorization and unlimited usage of materials including longer time usage with a principle of "cradle-to cradle" unlike "cradle-to-grave" type of scenario. In upcycling, field "chemical recycling" is attracting attention but faces limited commercial success. Upcycling through composite fabrication for new industrial uses shows promise. Although many argue pyrolysis of waste plastic is another strategy towards upcycling waste plastics, it is still considered at its niche stage of development. The whole upcycling strategy needs extensive research to innovate maximum value and high performance product of economic value to make it attractive and truly sustainable.

Acknowledgements: i) The Ontario Ministry of Agriculture, Food and Rural Affairs (OMAFRA) – University of Guelph, the Bioeconomy Industrial Uses Research Program Theme (Project Nos. 030486, 030578 and 030671); ii) OMAFRA-University of Guelph Gryphon's Leading to the Accelerated Adoption of Innovative Research (LAAIR) Program (Project No. 030416); iii) the OMAFRA- Ontario Agri-Food Research Initiative (Project No. 055217); iv) the Ontario Research Fund, Research Excellence Program; Round-9 (ORF-RE09) from the Ontario Ministry of Economic Development, Job Creation and Trade, Canada (Project Nos. 053970 and 054345); and (iv) the Natural Sciences and Engineering Research Council of Canada (NSERC), Discovery Grants (Project No. 401111).

667e Refined Chemicals and Carbonaceous Materials Derived from Biomass Pyrolysis

Yaseen Elkasabi

Global economic and environmental conditions have prompted the pursuit of alternatives to petroleum and coal-derived fuels and chemicals. The transition from fossil fuels to biofuels has relatively clear paths; however, a transition to biobased chemicals contains significantly more ambiguities, since specific pathways to produce many critical "drop-in chemicals" or "drop-in solvents" are unknown. Fast pyrolysis of biomass, when optimized, produces a bio-oil amenable towards many different separations and refining processes. In this talk, I will discuss recent advances in downstream refining of pyrolysis bio-oils. Advanced pyrolysis processes produce oils with sufficient quality ($O < 15$ wt%) for use in separations processes that were otherwise infeasible. Our laboratory and bench-scale continuous processes for bio-oil separations, such as distillation and extraction, will be reviewed. Second, the distillates can convert into fuels directly with simple catalysts and/or minimal processing. Third, when distilled, the bio-oils produce a solid residue that can be further heat treated into calcined coke, an electrically conductive material that is of critical use by the aluminum industry. I will review our lab's progress in both material quality improvement and process design and scale-up.

121a Synthesis of Paracetamol and 4-Aminophenol from Lignin-Derived Hydroquinone

Jimin Park, Marta Hatzell, Carsten Sievers, Andreas Bommarius

The active pharmaceutical ingredient (API) paracetamol, contained in over-the-counter (OTC) drugs such as Tylenol®, is one of the most widely used pharmaceuticals worldwide. Supply chain shortages due to COVID-19 have demonstrated the reliance of many nations on imports for their paracetamol supplies. These factors have raised interest in developing new methods of synthesizing paracetamol.

An important bottleneck to the synthesis of paracetamol is access to 4-aminophenol as an intermediate. While the acetylation of 4-aminophenol to paracetamol is selective and fast, acquiring 4-aminophenol is a significant challenge. The traditional synthesis of 4-aminophenol through the nitration of benzene is environmentally unfriendly because of factors such as high acid/salt waste and a reliance on petrochemicals for the feedstock. Lignocellulosic biomass offers an attractive route to paracetamol. Hydroquinone is a promising starting point for the synthesis of 4-aminophenol because it can be accessed through phenol, a significant product of lignin depolymerization. Aminating hydroquinone to form 4-aminophenol is a promising route that may improve Green Chemistry metrics for synthesizing this important intermediate.

A key difficulty to the synthesis of 4-aminophenol from hydroquinone is deciding the optimal route. We have examined the performance of several amination agents such as ammonium hydroxide, hydrazine monohydrate, ammonium chloride, and ammonium acetate regarding the direct amination of hydroquinone, targeting values such as conversion, yield, and selectivity. Choosing an ideal amination agent is not straightforward, as the amination step is rather slow and unselective, requiring high temperatures and long reaction times. A preliminary reaction mechanism suggests the importance of protonating hydroquinone to form a good leaving group for nucleophilic attack by the aminating agent. As such, we have examined the effect of various acidic catalysts such as zeolites, metal phosphates, metal oxides, and heteropoly acids, as well as the effect of pH, on the reaction conditions. A secondary issue is the stability of 4-aminophenol: under the reaction conditions, 4-aminophenol is prone to further reaction, acting as an additional nucleophile and forming high molecular weight compounds as side products. For that reason, we will also discuss the viability of a one pot synthesis, in which amination and acetylation are done simultaneously, with the acetylating agent "shielding" 4-aminophenol from further reaction to various side products.

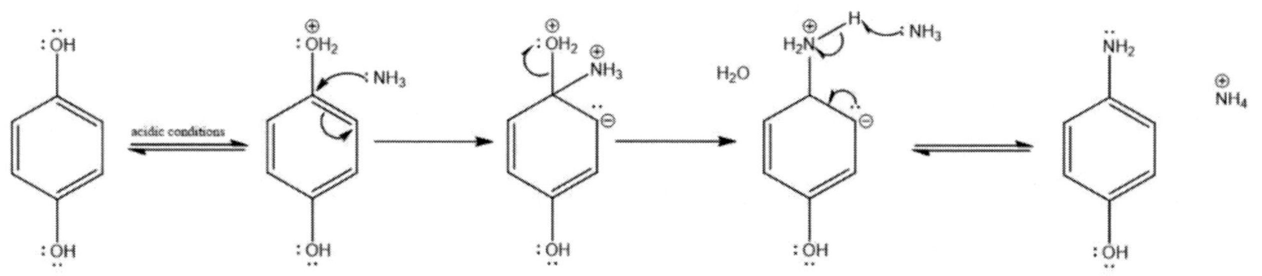

Figure 1. Preliminary mechanism for the amination of hydroquinone

121b Properties of Activated Carbon from Two Lignin Waste Streams in Lignocellulosic Biorefineries

Chengjun Wu, Graham W. Tindall, Carter Fitzgerald, Mark C. Thies, Mark E. Roberts

Creating fuels and chemicals from lignocellulosic biomass offers a promising alternative to fossil fuels for accomplishing carbon neutrality and minimizing greenhouse gas emissions. To economically produce the fuels from biomass, it is essential to convert the "ligno-" components to valuable products. Activated carbon (AC) is a suitable product because it's market is large enough to consume the amount of lignin separated during the biofuel production and it does not have a stringent requirement for its precursor. Currently, the two main lignin streams from biorefineries are the black liquor produced during alkaline pre-treatment and solids remaining after enzymatic digestion.

In this work, we use the solid contents of the two main lignin streams from biorefineries as precursors to produce AC. These streams are briefly separated and purified to create solid fractions with reduce ash and enriched with lignin. Solid fractions with varying ash (impurities), lignin content, and lignin properties were used as precursors for AC to identify how the precursor properties affect AC properties. The pore structure was analyzed using BET analysis and the aqueous adsorption capacities were characterized by using iodine number and methylene blue value. Using standard $ZnCl_2$ activation and low-temperature carbonization, we synthesized ACs with surface areas up to 2500 m^2/g, which can likely be increased by optimizing the synthesis conditions. Importantly, we found that ash content had a small affect and sugar content had a strong influence on the AC pore properties. Furthermore, the proposed method for synthesizing AC is not complicated and is appropriate for bulk production.

121c Biobased Composites for Improved Mechanical Performance and Recyclability

Xianglan Bai, Yixin Luo, Moham Razzaq, Baker Kuehl, Eric Cochran

In this study, the mechanical properties of polylactic acid (PLA) and PLA-based composites were reinforced with lignin based nanocarbons synthesized using Joule-heating method. Three different blending methods were investigated, which include direct blending of PLA and nanocarbons, pre-mixing PLA and soybean based hyperbranched elastomers (namely SBCP-B) followed by blending the PLA composite with nanocarbons, and one-step blending of PLA, soybean based hyperbranched elastomers and nanocarbons. The effect of fabrication condition of nanocarbons and the mass loading of nanocarbons in composites were investigated. The analysis results showed that blending 0.1wt% of nanocarbons could increase tensile strength of PLA from 75.96 MPa to 85.98 MPa without affecting tensile modulus and strain of the material. Adding 0.5% of nanocarbons enhanced the performance of SBCP-B by increasing tensile strength from 68.72 MPa to 71.78 MPa, tensile modulus from 5.43 MPa to 5.59 MPa, and strain from 48.59% to 55.91%. When PLA, soybean based hyperbranched elastomers, and 0.5%load nanocarbons were blended in one batch, tensile strength of 82.61 MPa, tensile modulus of 6.08 GPa, and strain of 7.72% were obtained.

121d Sustainable Green Composite Materials from Advanced Biocarbons: A Path Forward to Circular Bioeconomy

Manjusri Misra

Lightweight green composites from renewable resources are attracting increased attention from the advanced manufacturing sectors recently. A strong momentum for weight reduction in auto parts is motivated by a need for better fuel economy. To increase renewable content in materials, we use thermo-chemical conversion (pyrolysis) of biomass/agri-food residues to overcome the key challenges associated with the use of traditional natural fibrers in melt processed biocomposites. Biocarbons obtained from inexpensive biomass, including farm wastes/residues, food processing wastes, and lignin show tremendous potential for affordable biocomposite materials. In addition, biocarbons have also been used effectively in composites with waste plastics. Overall, biocarbon composites can be very effective as one of the enabling strategies towards sustainable materials and a circular bioeconomy.

Acknowledgement: (i) the Ontario Research Fund, Research Excellence Program; Round 9 (ORF-RE09) Ontario Ministry of Economic Development, Job Creation and Trade (Project Nos. 053970 and 054345); (ii) the Ontario Ministry of Agriculture, Food and Rural Affairs (OMAFRA)/University of Guelph – Bioeconomy for Industrial Uses Research Program (Project Nos. 030331, 030332); (iii) the Natural Sciences and Engineering Research Council of Canada (NSERC), Canada Research Chair (CRC) program Project No. 460788; and (iv) The NSERC,-Collaborative Research and Development Grants (CRD) Project No. 401637 with the partner industries Prism Farms Limited and Competitive Green Technologies, Lamington, Ontario, Canada.

534b Application of Protocatechuic Acid-Based Deep Eutectic Solvent for Utiliztion of Engineered Biomass

Yunxuan Wang, Anqi Ji, Xianzhi Meng, Yang Tian, Linjing Jia, Aymerick Eudes, Kwang Ho Kim, Yunqiao Pu, Gyu Leem, Deepak Kumar, Jeong Jae Wie, Arthur Ragauskas, Chang Geun Yoo

To efficiently valorize lignin and achieve a sustainable biorefinery process, genetically engineered sorghum bagasse with altered lignin monomers was pretreated by a novel protocatechuic acid-based deep eutectic solvent (DES). We synthesized the DESs with protocatechuic acid (PCA) and choline chloride (ChCl) and applied this solvent to fractionate sorghum bagasse. The transgenic sorghum bagasse having protocatechuate incooperated lignin was used as a feedstock in this study. Additionally, three-component DESs containing ChCl, PCA and water or ethylene glycol were formed and tested. The fractionated lignin from the black liquor was subjected to hydrothermal treatment to recover protocatechuic acid, and the cellulose-rich pulp was enzymatically hydrolyzed. Structures of isolated lignin before and after hydrothermal treatment were characterized by two-dimensional heteronuclear single quantum (2D HSQC) NMR and gel permeation chromatography, while the liquid products of hydrothermal treatment were analyzed by gas chromatography-mass spectrometry (GC-MS).

534c Effects of pH and Multistage ALPHA Process on the Phase Behavior of Kraft Lignin

Oreoluwa Agede, Graham W. Tindall, Mark C. Thies

As a key component of lignocellulosic biomass, lignin is an abundant biopolymer, notable for its relatively high amount of carbon, aromatic moieties, and functional hydroxyl groups. Compared to polysaccharides, lignin can achieve higher char yields, benefiting applications such as activated carbon or carbon fibers. Furthermore, the aromatic hydroxyl groups, and other specific chemistry not commonly found in other biomolecules open pathways for lignin-based polyurethanes, phenol-formaldehyde resins, or polyesters. However, as a consequence of its biomass origins, lignin applications must also contend with the impure and polydisperse nature of this resource. Extraction methods such as the Kraft pulping process, or the alkaline pulping used by prospective bioethanol refineries, do not target lignin as the primary output. Instead, lignin is often treated as a co-product at best, and a by-product at worst. This results in lignin streams regularly containing as much as 5% ash, 5% sugar, and varying extents of chemical modification (sulfur addition in the case of Kraft lignin).

With renewability and potential for low-cost underpinning all things lignin, these paradigms must be upheld in all downstream fractionation and purification. To meet this end, a new process of fractionating lignin was developed in 2016 by Thies and coworkers called Aqueous Lignin Purification with Hot Agents (ALPHA). This process makes use of a mixture containing an organic solvent such (acetic acid, ethanol, etc.) and water to form two liquid phases. These phases consist of a lignin-rich phase with high molecular weight and high purity, along with a solvent-rich phase with low molecular weight and lower purity. This method of purifying lignin has proven to be successful with previous studies showing an ultrapure fraction with <25ppm of sodium, and controlled high molecular weight of >10,000 Da.

To optimize the ALPHA process and achieve control of lignin purity and molecular weight, the underlying phase behavior must be understood. There are several operating conditions to consider: organic solvent concentration, temperature, solvent-to-feed ratio, and pH. While the effects of manipulating the temperature, organic solvent, and solvent-to-feed ratio have been previously studied, the effect of system pH has not. It has since been found that pH has a significant impact on phase behavior – suggesting that (de)protonation of lignin is likely to impact its chemical properties. When dealing with lignins coming from different processes, with different levels of acidic/basic impurities, understanding the role of pH in the system is critical for successful fractionation.

In this work, the role of pH and molecular weight on the phase behavior of selected lignins are investigated. This knowledge enables a multi-stage process that can successfully control the molecular weight and purity of lignin fractions in order to enable down-stream applications research.

695a Simultaneous Biomass Fractionation and Xylan Conversion Using Deep Eutectic Solvents

Caixia Wan, Qianwei Li

A biphasic system consisting of DES pretreatment and organic solvent was used for simultaneous switchgrass fractionation and conversion. High furfural yield was achieved from the one-pot reaction. The solid residue was rich in cellulose, which was readily processed into fermentable sugars and nanocellulose. Lignin was recovered with a high purity, showing low molecular weight and good reactivity. The solvent system also showed good recyclability and reusability. This study provided a novel solvent system and approach for lignocellulose fractionation and direct conversion of furfural from lignocellulose.

695b Reinforced Learning Based Control Algorithm for Anarobic Digestion Under Feedstock Uncertaint

Abigael Wahlen, Ji Gao, Caleb Ju, Guanghui Lan, Zhaohui Tong

Essential to creating a sustainable future is implementing and optimizing alternative technologies that can help mitigate the current carbon crisis. However, many biological systems are plagued by uncertainty due to the complexity of organic feedstocks. Anaerobic digestion systems experience this overriding uncertainty in the feedstock, especially when multiple feedstocks are present. A reinforced learning algorithm can be used to control and maintain this process to solve this issue. First, several hundred data points were collected from literature to use in this algorithm. Then, due to inconsistencies in feedstock analysis throughout literature, a linear optimization algorithm was used to determine the biomolecular composition of the feedstocks to implement in an anaerobic digestion model. From this data, random samples were taken to establish the uncertainty in our model. To control this process, two different learning algorithms are analyzed for comparison. A Markov decision process (MDP) that optimizes and controls anaerobic digestion under the established uncertainty. Also, a reinforced learning algorithm where the uncertainty is not established. Next, these algorithms are implemented into a kinetic model based on anaerobic digestion model no. 1(ADM1) to determine the target methane output for our process. To apply these algorithms, ADM1 is simulated, determining the efficiency under several conditions. These conditions are a steady-state model, a step model, and a continuously changing model. This model will allow for an alternative automated control system for this delicate process, ensuring stability in the reactor. Furthermore, these results will lead to understanding how scientists and engineers can apply reinforced learning to complex biological systems like anaerobic digestion.

695c Renewable Tar Pitch Based on Pyrolysis Bio-Oils.

Yaseen Elkasabi, Charles A. Mullen

Over the years, fossil-based carbonaceous solids have decreased in quality, with respect to their use in industrial applications like aluminum smelting. Specifically, coal tar pitch is a binder material for smelting anodes. As the quality of coal has strayed from the norm, so has the supply of and critical properties of pitch, such as toluene insoluble (%TI), coking value, etc. Furthermore, a biobased alternative to coal tar pitch will certainly aid in reducing nonrenewable carbon emissions from the manufacturing industry. This talk will discuss an approach towards producing renewable tar pitch from pyrolysis oils. Partially deoxygenated pyrolysis oils underwent distillation, and the solid residues subsequently underwent extraction of toluene solubles. This soluble fraction underwent heat treatment steps to produce pitch with amenable coking values (>43 wt%) and toluene insoluble fractions (32 – 54 wt%). The toluene insoluble fractions of the residues were found to consist primarily of densely oxygenated compounds like levoglucosan.

695e Experimental Analysis of Convective Drying of Paper and Board

Koushik Sampath, Leonard Reynolds, Michael Ringold, Xinyi Li, Huajiang Huang, Sridharan Ramaswamy

In conventional multi cylinder drying of paper and board both conductive drying from steam heated dryer cylinders and convective drying by flowing air over the surface in the pockets are used. Pocket ventilation is a critical component in providing additional thermal energy and more importantly carrying away the evaporated water. Air temperature, velocity and humidity are critical process parameters influencing the convective drying process. An experimental setup was developed to study the convective drying of paper and board. Paper sheet moisture, temperature as well as air velocity, temperature and humidity were measured as drying proceeds. The instantaneous drying rate as well as convective heat and mass transfer coefficients were determined as drying proceeds. Experimental results as well as comparison to literature and commercial data are presented. The convective heat and mass transfer coefficients determined were compared to traditional correlations normally used in the modeling of paper drying. This data will be useful in process development and modeling and simulation of paper drying processes.

695f Process Evaluation of Lignocellulosic Resource Treatment for Sustainable Corrosion Inhibitor of Carbon Steel

Tianxing Cai

Lignocellulosic resource is the biomass material with polysaccharides and aromatic polymers. This biosourced chemical and material can be capable of producing value ended products such as sustainable corrosion inhibitors. The purpose of this study is to find the relationship between process condition to handle lignocellulosic resource and the anti-corrosion performance under the acidic and salty environment. Since the corrosion is still one of the industrial challenges to the equipment, pipeline and other infrastructure, the results from this study will benefit the environment, chemical industry, and give contribution to knowledge in the field.

598a Utilization of Marine Macroalgae for the Production of Biofuels and Biomaterials Via Hydrothermal Pyrolysis Process

Justinus Satrio, Apip Amrullah, Obie Farobie, Michael Berzolla, Hamed Bazrafshan

Algae are known as the third-generation feedstock which are highly promising to be used for the synthesis of bio-based fuels, energy, and materials via biochemical, thermochemical, and biological conversion techniques. Compared to land biomass, algae have the advantages of having faster growth rates, in addition to potentially much higher yields per unit surface area of land. Furthermore, algae are unique since they contain carbohydrates, lipids, and proteins, which make them more potential for producing wider range of products via various processes. A significant shortcoming of algae as biofuel feedstock is the need of dewatering and drying processes which are both energy- and cost-intensive. Eliminating dewatering and drying pre-processing, prior to the conversion process, will be a significant cost and energy reduction.

In this study, three distinct species of marine macroalgae, i.e., *Euchema cottonii*, *Sargassum sp*, and *Ulva lactuca,* were used as feedstock for the synthesis of hydrochar and organic liquid via hydrothermal pyrolysis (HTP) process. HTP process is a thermochemical process that converts wet biomass materials to a solid, liquid, and gas products. Under sub-critical conditions, water in the HTP reactor stays in liquid form and acts as a reaction medium to promote the breakdown and cleavage of chemical bonds in the solid biomass. The absence of liquid-to-vapor phase change of the water makes the process significantly less energy intensive compared to a process that involves water vaporization or drying. The temperature range for HTP process is from 180-370°C. At lower reaction temperature range (180–240°C), known as hydrothermal carbonization (HTC), the main product is hydrochar, which is the solid product. The HTP process at higher temperature range, known as hydrothermal liquefaction (HTL), will produce liquid, i.e., bio-oil, as the primary product.

The focus of this research was to evaluate the effects of feedstock and HTP process parameters on the yields and selectivity, along with physicochemical properties, of the solid and liquid products. HTC reaction temperature and retention time were the major parameters that were varied. Retention times varied from thirty minutes to two hours and the temperature range was focused between 200°C and 300°C.

Various analytical and characterization testing will be conducted on the liquid and solid products of the HTP process. The characterization techniques will include CHNOS elemental analysis, approximate analysis (TGA), specific surface area, porosity, heating value, and microstructure (SEM analysis) of the hydrochar products. The liquid analysis to evaluate chemical compositions will be performed by using gas chromatography techniques. Specific attention will be given on the fate and the recovery of the lipid component. When the desired product is hydrochar, the lipid component in algae feedstock will remain in the solid product. Compared to extracting lipid from the original algae feedstock, extracting lipid from algae-derived hydrochar is significantly much easier due the "opening" of the solid structures. Reduction of the energy and cost requirements of extracting lipid from algae will be a noteworthy progress toward the commercial utilization of algae as renewable feedstock for producing biofuels and materials.

CHEMICAL PRE-LITHIATION OF LIGNIN-DERIVED HARD CARBON AIMED FOR LITHIUM-ION BATTERY ANODE WITH HIGH-RATE PERFORMANCE

Ruiqing Zhang[a], Xiaodong Hou[a], Xin Zhang[a], Yun Ji[a,]*

a. University of North Dakota, Grand Forks, ND, 58202.

Abstract

Recently, the environmental problems, the reduction of fossil fuel sources, and the severe environmental safety policies are forcing the battery industry to be sustainable. In the energy industry, there is a great interest in developing more efficient energy storage devices using renewable raw materials.[1] Because lignin is renewable and abundant, lignin-derived hierarchical porous carbons have been widely studied as promising electrode materials for energy storage. The use of chemical activation agents causes lignin to release carbon, hydrogen, and oxygen atoms, forming porous structures in lignin macromolecules, thus enhance the lithium storage and adsorption capacity. However, these pore structures also reduce the initial Coulombic efficiency (ICE) significantly. This large initial capacity loss results in high lithium ions consumption in the electrolyte and needs to be compensated by excess amount of cathode material, posing severe challenges for practical battery applications. New technology is needed to solve the problem of low ICE of lignin-based electrode materials.

In this study, pre-lithiation through the spontaneous chemical reaction between the aryl etherification reagent was used to improve the ICE of the lignin-based negative electrode material. Our results showed that the ICE of the activated lignin-based negative electrode material was increased from 48.9% to 82.7%. The results indicate that treated lignin is a promising renewable electrode material for battery applications due to its high yield, low price, high ICE, high capacity, and stable cycling. It can be used in electrochemical energy storage and conversion devices, such as supercapacitors, fuel cells, lithium-ion batteries, etc.

1. Introduction

Rechargeable lithium-ion batteries (LIBs) are the most widely used energy storage devices for portable electronics and electrical vehicles. Although widely adopted in the vehicle market, lithium-ion batteries still require further development to sustain their dominating roles among competitors.[2,3] For example, an important part of LIBs, anode materials still have many deficiencies that need to be improved. Among various materials for LIB anodes, carbonaceous anodes including natural/synthetic graphite and soft/hard carbons, represent the most practically viable candidates for LIBs because of their diverse structures and properties as well as low cost and widespread availability.[4,5] In contrast, hard carbon (HC), a non-graphitic carbon, seems to be a better option as a high capacity and high-power anode for future LIB technologies.[6]

The performance of hard carbon relies primarily on the precursors and the synthesis approaches. Lignin, as the second-abundant biopolymer in nature, with the low cost and high carbon yield makes it an ideal precursor for the preparation of hard carbon.[7] But lignin also faces the same problems as other hard carbon materials, such as low initial Coulombic efficiency (also known as ICE) and low rate performance due to structural disorder.

To address this challenging problem, various pre-lithiation strategies[8-11] have been developed to pre-lithiate the HC anode for compensating the loss of Li ions in the initial cycles and thereby upgrading the electrochemical performance of HC anode. Compared with many difficult and expensive pre-lithiation operations, a simple and low-cost pre-lithiation method is needed.

Here, we report a facile and efficient pre-lithiation approach to prepare pre-lithiated hard carbon (PHC) by chemical reaction of HC with a lithiation reagent (Lithium Naphthalenide, L-N). After pre-lithiation reagent treatment, the solid electrolyte interphase (SEI) layer formed on the PHC anode after the chemical pre-lithiation process is thinner, denser, and more robust than those grown on pristine HC anode during normal charge/discharge cycles, thus enabling a very high ICE, high power, and stable cyclability of the PHC anode. Based on this PHC anode, our assembled half-cells exhibit a high ICE of 82.7% at 50mAh/g and a capacity retention close to 100% over 700 cycles, demonstrating a competitive advantage over similar LIBs using graphite anodes.

2. Experimental Section

2.1 Materials synthesis

The dried lignin powder was put into a tube furnace and carbonized at 800 °C for 2 h at a heating rate of 5 °C min^{-1} under nitrogen flow. Then, the powder was ground for 30 min and sieved, and the final sample was named CL. The pre-lithiation process was performed by simply immersing the HC electrode in 1 mL of 1 mol L^{-1} naphthalene lithium solution in a vacuum drying oven before assembling the battery. Afterwards, the PHC electrode was washed 2 times with DME and dried in a vacuum chamber for 10 min. DME was used as a solvent because it is compatible with Li naphthalenide reagent.

2.2 Materials characterization

The morphology of the sample was observed by scanning electron microscopy (SEM, FEG-650, U.S.) and the carbon contents of the samples were measured by Total Organic Carbon (TOG, ASI-V, Japan).

2.3 Electrochemical measurements

The electrochemical electrode was prepared as explained below. 80 wt% active materials (CL), 10 wt% acetylene black and 10 wt% carboxymethyl cellulose (CMC) were mixed in aqueous solution to form a uniform slurry and then casted on Cu foil. The electrode was dried at 80 °C in vacuum oven for 12 h and punched into circular electrode before use. The loading mass of the CL is controlled around 3.0 mg cm^{-2}. An electrolyte of 1.0M $NaPF_6$ in Ethyl Methyl Carbonate (EMC)/dimethyl carbonate (DMC)/ethylene carbonate (EC) (1/1/1 by volume) with 2% fluoroethylene carbonate (FEC) was used therein. Cells were assembled in an argon-filled glove box and evaluated over a voltage range of 0–3.0 V at 25 °C using the Neware battery test system (China). Cyclic voltammetry was measured on the Interface 1010E General-purpose electrochemical workstation (U.S.) at a scan rate of 0.1mV/s in the voltage range from 3.0 to 0.01V (vs. Li+/Li and hereafter). Alternating current electrochemical impedance spectroscopy (EIS) was measured in the frequency range from 100 kHz to 100 MHz with an alternating current voltage amplitude of 10 mV at the open-circuit voltage. All electrochemical measurements were performed using CR-2032 type coin cells.

3. Result and Discussion

The synthesis and redox behavior of L-N was selected as an efficient and widely applicable pre-lithiation agent. As shown in Fig. 1 when the circular electrode piece is soaked in the pre-lithiation solution, the HC undergoes a chemical reduction reaction and electrons are transferred from naphthalene anion to Li^+ and simultaneously inserted into the graphite layer of HC to form lithiated host (PHC). Once the PHC is formed, its low potential initiates the reductive decomposition of dimethylether (DME) solvent to form a porous and immature SEI layer that consists of organic composition (ROLi). This immature SEI layer would be further strengthened when contacted with the Li-ion battery electrolyte, where FEC and PF_6^- anions decompose to produce LiF^- rich SEI layer, finally leading to a dense and robust SEI film.[6]

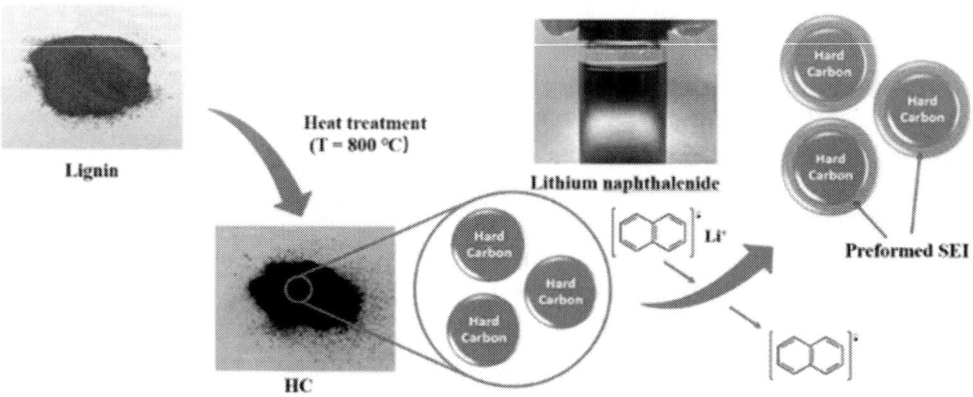

Figure. 1 Schematic illustration of the chemical pre-lithiation reaction of HC in Li-Naphthalenide/DME solution and SEI formation on the PHC anode

Figure 2a and 2b show the SEM images of the treated and untreated electrode materials after cycling. We can find that the SEI film on the surface of the pre-lithiated anode disk is denser and more uniform. In contrast, the SEI distribution on the surface of the anode disk without pre-lithiation, which only spontaneously generates an SEI film through battery cycling, is very uneven, and a part of the hard carbon material is not covered by the SEI film and is directly exposed to the electrolyte. The uneven distribution of SEI may have an adverse effect on the electrochemical performance of the battery.

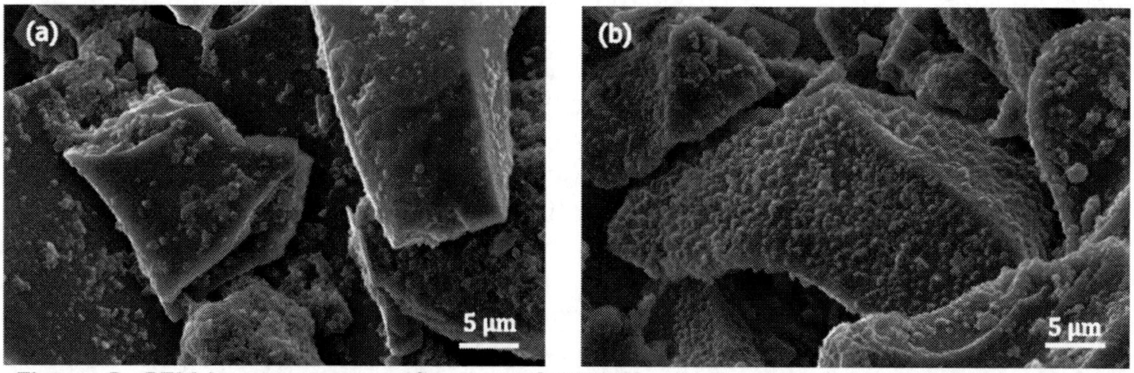

Figure 2. SEM image at magnification of 500X for lignin-derived hard carbon anode electrode (a) Electrode without pre-lithiation (b) Electrode with pre-lithiation.

To optimize the pre-lithiation degree, we compared the charge/discharge behaviors and the ICE values of the PHC anodes obtained at different pre-lithiation time. As shown in Figure 3a) and 3b), the PHC anode exhibited significantly lowered discharge (lithiation) plateaus and greatly decreased initial irreversible capacities with increased pre-lithiation time. After 3 hours of pre-lithiation, the PHC anode exhibited a sufficiently high ICE value above 81%, which means that the initial irreversible capacity has been greatly improved by a full 32% from the original 49%, which shows that pre-lithiation can significantly improve the Initial Coulombic Efficiency of lignin-derived hard carbon.

The rate capabilities of both the HC and PHC anodes were comparatively evaluated by galvanostatic charge/discharge at varying current rates from 100 to 2000 mA g−1. As shown in Figure 3c, the PHC anode demonstrates an improved rate performance with reversible capacities of 178, 130 and 61 mAh g^{-1} at high rates of 500, 1000, and 2000 mA g−1, respectively. In comparison, pristine HC anode delivers much lower capacities of 130, 75, and 27 mAh g−1 at the current rates of 5000, 1000, and 2000 mA g−1, respectively. This significant difference in the rate capability suggests that the chemical pre-lithiation may produce a better SEI film than conventional electrochemical cycling pre-lithiation on HC anodes, thus leading an improved high rate performance.

The better SEI film produced by chemical pre-lithiation can also be evidenced from the cycling performances of the HC and PHC electrodes. As displayed in Figure 3d, HC anode delivers 293.0 mAh g^{-1} at the current of 200 mA g−1 over 700 cycles, Nevertheless, the PHC anode can not only realize a high reversible capacity of 296 mAh g^{-1}over 500 cycles at 200 mA g−1 but also display a high capacity retention close to 100% over 700 cycles, mainly due to the higher stability of the SEI layer on the PHC anode.

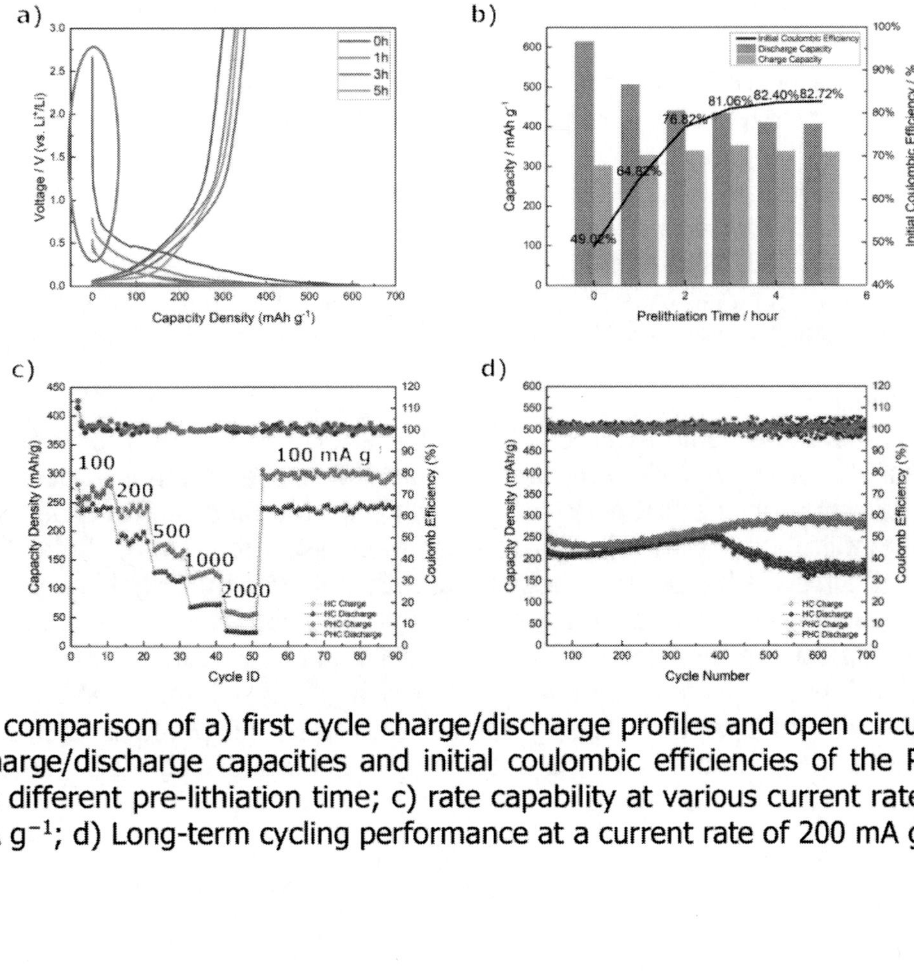

Figure 3. A comparison of a) first cycle charge/discharge profiles and open circuit voltages; b) initial charge/discharge capacities and initial coulombic efficiencies of the PHC anodes obtained at different pre-lithiation time; c) rate capability at various current rates from 100 to 2000 mA g^{-1}; d) Long-term cycling performance at a current rate of 200 mA g^{-1}.

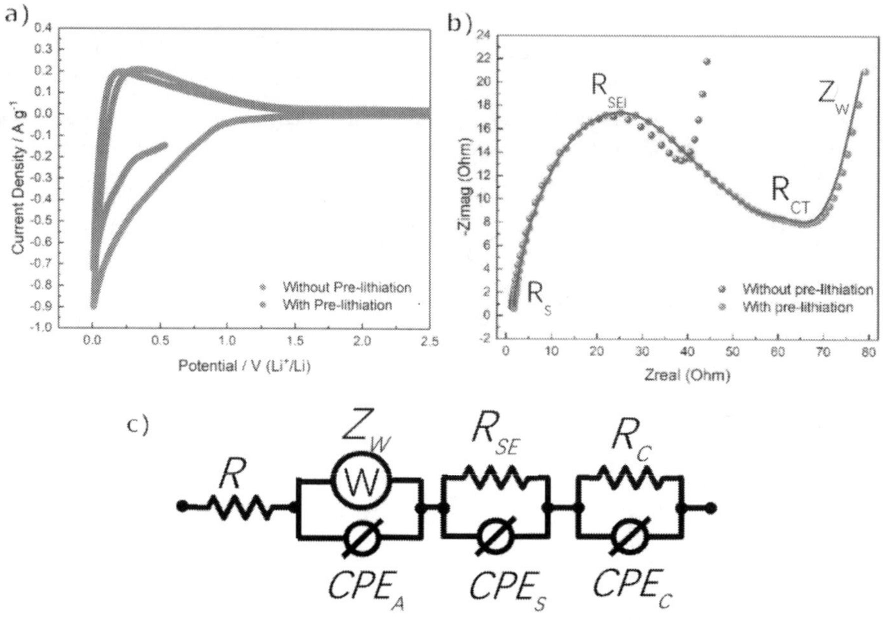

Figure 4. a) CV curves b) Nyquist plot of HC and PHC measured at the open circuit of HC/Li and PHC/Li half-cell; (c) Equivalent circuit diagram corresponding to the EIS curves.

Figure 4a) shows the cyclic voltammetric (CV) curves of the HC anode and PHC anode at the first cycles. We can see that the initial voltage of the pre-lithiated battery drops to 0.5 watt, suggesting that there has been an SEI film already existing on the surface of the anode and the preformed SEI functions well for Li+ insertion reaction. We also conducted a kinetic analysis for fresh anodes. Electrochemical impedance spectroscopy (EIS) is a tool used for mechanistic and kinetic analysis inside the lithium-ion battery. Figure 4b) illustrates the Nyquist plots of EIS spectra and the equivalent circuit model. According to the EIS, we can build an equivalent circuit for analyzing the resistance of every part of the lithium-ion battery. R-S, the part of the impedance curve close to 0 ohms is the ohmic series resistance associated with the cell components. R-SEI, a large semicircle, represents the Li-ion transfer resistance through the SEI layer. R-CT, a small semicircle, is the charge transfer resistance.
Z-W, like a straight line, is the Lithium Ion Diffusion Impedance, associated with the Li-ion diffusion into the active electrode. From here, we can see that pre-lithiated anode shows an SEI semi-cycle. Which means a SEI layer was generated in a fresh pre-lithiated anode.

4. Conclusion

In summary, we demonstrated a facile and effective approach to obtain a PHC by a chemical reaction of HC with lithium naphthalenide. This pre-lithiation process generates a preformed SEI layer on the PHC electrode, which then transforms into a dense and robust SEI film upon infiltration into the electrolyte after initial discharge. This SEI film can effectively compensate the irreversible capacity loss and greatly enhance the initial and cycle Coulombic efficiencies, leading to high rate performance and cycle stability of PHC anodes. Based on the PHC anode, we assembled PHC/Li half cells and evaluated their electrochemical performances. The PHC/Li cells demonstrated very high ICE values of >81.0%, and a very high stability with little capacity fading after 700 cycles. The practical application of pre-lithiated lignin-derived hard carbon materials in next-generation high-capacity and high rate performance LIBs is demonstrated.

Acknowledgement

The authors appreciate the support from the University of North Dakota's Department of Chemical Engineering and Institute for Energy Studies.

Conflict of Interest

The authors declare no conflict of interest.

Keywords

Chemical pre-lithiation, Lignin, Hard carbon, Li-ion battery, Lithium Naphthylamide.

References

1. Espinoza-Acosta, José Luis, et al. "Lignin in storage and renewable energy applications: A review." *Journal of energy chemistry*, 2018, 27(5): 1422-1438.
2. Liu, Dongqing, et al. "Review of recent development of in situ/operando characterization techniques for lithium battery research." *Advanced Materials*, 2019, 31(28): 1806620.
3. Zeng, Xiaoqiao, et al. "Commercialization of lithium battery technologies for electric vehicles." *Advanced Energy Materials*, 2019, 9(27): 1900161.
4. Dou, Xinwei, et al. "Hard carbons for sodium-ion batteries: Structure, analysis, sustainability, and electrochemistry." *Materials Today*, 2019, 23: 87-104.
5. Xie, Lijing, et al. "Hard Carbon Anodes for Next-Generation Li-Ion Batteries: Review and Perspective." *Advanced Energy Materials*, 2021, 11(38): 2101650.
6. Shen, Yifei, et al. "Chemically pre-lithiated hard-carbon anode for high power and high capacity Li-ion batteries." *Small*, 2020, 16(7): 1907602.
7. Lin, Xiuyi, et al. "Advanced lignin-derived hard carbon for Na-ion batteries and a comparison with Li and K ion storage." *Carbon*, 2020, 157: 316-323.
8. Zhao, Jie, et al. "A general prelithiation approach for group IV elements and corresponding oxides." *Energy Storage Materials*, 2018, 10: 275-281.
9. Xu, Hui, et al. "Roll-to-roll prelithiation of Sn foil anode suppresses gassing and enables stable full-cell cycling of lithium-ion batteries." *Energy & Environmental Science*, 2019, 12(10): 2991-3000.
10. Wang, Gongwei, et al. "Chemical prelithiation of negative electrodes in ambient air for advanced lithium-ion batteries." *ACS applied materials & interfaces*, 2019, 11(9): 8699-8703.
11. Liu, Xiaoxiao, et al. "A simple electrode-level chemical presodiation route by solution spraying to improve the energy density of sodium-ion batteries." *Advanced Functional Materials*, 2019, 29(50): 1903795.

24b Sustainable Biocomposites from Walnut Shell Powder Filled Poly(Butylene Succinate-co-Butylene Adipate) (PBSA)/Poly(Butylene Adipate-co-Terephthalate) (PBAT) Blend

Dayna McNeill, Akhilesh Pal, Amar K. Mohanty, Manjusri Misra

Traditional practices of production and consumption of petroleum-based plastic for packaging applications are becoming obsolete due to the negative implications of non-biodegradable plastics on the environment. The current research investigates the utilization of home compostable polymers with discarded food waste as a feasible and sustainable alternative in packaging applications. PBAT comprises aromatic and aliphatic units, exhibits higher toughness and %elongation at break, making it suitable for various applications. PBSA, derived from polybutylene succinate (PBS), exhibits significant variability for several processing methodologies and good mechanical properties such as flexibility. Walnut shells (WS) are used as a natural filler due to their tough exterior and chemical composition (49.7% polysaccharides, 30.1 % lignin and 10.6% extractives). The previous research has primarily focused on polymer blends with polybutylene adipate terephthalate (PBAT). However, limited research has been performed on polymer blends, including polybutylene succinate-co-adipate (PBSA) and PBAT. This study will investigate the effectiveness of such a blend comprising PBSA:PBAT (60:40) in the presence of waste walnut shells powder (WSP)in a range of 15 – 50 wt%. Incorporating a higher content (up to 50 wt%) of WSP with biopolymers blend through the injection moulding process, a relatively higher heat deflection temperature (HDT), i.e., 79 °C, was observed as compared to PBSA/PBAT blend (53 °C). The tensile and flexural modulus of the PBSA/PBAT blend with WSP was improved up to 608.40±60.80 and 549.80±27.97 MPa as compared to PBSA/PBAT blends. Such improvements in modulus and HDT were observed due to the flakey fibrous structure of WSP. The impact strength of biocomposite was determined as 48.03±1.74 J/m, which is sufficient for rigid packaging applications. Incorporating a compatibilizer in biocomposites can significantly improve the interfacial adhesion between the components, which helps to enhance the mechanical and thermal properties. It is noteworthy that adding high filler content in the polymer blends reduces the overall cost of the final products (e.g., utensils and containers) and increases the valorization of waste agro-based residues in value-added applications.

Acknowledgements: (i) the Ontario Research Fund, Research Excellence Program; Round 9 (ORF-RE09) Ontario Ministry of Economic Development, Job Creation and Trade (Project Nos. 053970 and 054345); (ii) the Ontario Ministry of Agriculture, Food and Rural Affairs (OMAFRA)/University of Guelph – Bioeconomy for Industrial Uses Research Program (Project Nos. 030486, 030578); (iii) the Natural Sciences and Engineering Research Council of Canada (NSERC), Canada Research Chair (CRC) program Project No. 460788; and (iv) The NSERC Discovery Grants Project No. 401716.

24c Polybutylene Adipate Terephthalate-Based Biodegradable Composite Films: Effect of Talc Types on Mechanical and Water Vapor Barrier Properties

Shiv Shankar, Amar K. Mohanty, Manjusri Misra

The application of biodegradable and biobased plastics is a primary condition in the effective transition towards a circular economy. Plastic commodities primarily used in packaging should be biodegradable and recyclable to avoid the accumulation of plastic material in the environment. Research in the past few decades has led to the commercialization of a number of bio-based, biodegradable, and recyclable polymers. However, bio-based and biodegradable polymers typically have poor mechanical and barrier properties than the non-biodegradable plastic materials they intend to replace. However, this can be improved by adding mineral fillers. Therefore, the present study aims to develop polybutylene adipate terephthalate (PBAT)-based composite films reinforced with 20 wt% of three different types of talc (Jetfine 3CC talc, silane treated talc, and TAL-10000 talc). The morphology of various talc was observed using a scanning electron microscope. The effect of different talc on the mechanical and water barrier properties of PBAT-based films was studied. The addition of 20% talc increased the modulus of the composite films from 56.2 MPa to more than 160 MPa; however, elongation at break decreased significantly. The water vapor permeability of the neat PBAT film was 771.49 g.mil/m^2.day, which was decreased to 743.08, 615.52, 476.68 g.mil/m^2.day when 20 wt% TAL-10000 talc, silane treated talc, and Jetfine 3CC talc were added in the composite films, respectively. As the results show, the PBAT-based composite films with improved mechanical and barrier properties have the potential for packaging applications.

Acknowledgements: The authors would like to thank the following for their financial support: i) Ontario Ministry of Agriculture, Food and Rural Affairs (OMAFRA)/University of Guelph-Bioeconomy for Industrial Uses Research Program (Project Nos. 030486, 030578); ii) OMAFRA-Gryphon's Leading to the Accelerated Adoption of Innovative Research (LAAIR) Program (Project No. 030416); iii) OMAFRA- Ontario Agri-Food Research Initiative (Project No. 055217); iv) the Ontario Ministry of Economic Development, Job Creation and Trade ORF-RE09-078 (Project Nos. 053970, 054345); and v) the Natural Sciences and Engineering Research Council of Canada (NSERC), Canada Research Chair (CRC) program Project No. 460788.

24d Biodegradable Plastic Blends and Modified Starch-Based Sustainable Composite Films for Packaging Applications

Akhilesh Pal, Amar K. Mohanty, Manjusri Misra

Poly(butylene adipate-co-terephthalate) (PBAT), a well-known biodegradable random copolymer, was investigated with poly(3-hydroxybutyrate-co-hydroxyvalerate) (PHBV) as a biodegradable polymer with high barrier properties in the presence of fillers i.e. starch and silane treated starch. This was performed to determine the effect of silane-treated starch on the barrier, mechanical, and thermal properties of the cast melt extrusion of PBAT/PHBV blend. Significant improvements in oxygen (~91%) and water vapor (~82%) barrier properties of the PBAT/PHBV blend were analyzed after adding silane-treated starch as compared to PBAT film. This was due to the increased % crystallinity and uniform dispersion of modified starch in the blend matrix as confirmed by SEM analysis and the strong interfacial adhesion between filler and the blend matrix. The % elongation at break value for the cast extruded PBAT/PHBV film with modified starch was observed ~529%, which is suitable for flexible packaging applications.

Acknowledgements: The authors would like to thank the following for their financial support: i) Ontario Ministry of Agriculture, Food and Rural Affairs (OMAFRA)/University of Guelph-Bioeconomy for Industrial Uses Research Program (Project Nos. 030486, 030578); ii) OMAFRA-Gryphon's Leading to the Accelerated Adoption of Innovative Research (LAAIR) Program (Project No. 030416); iii) OMAFRA- Ontario Agri-Food Research Initiative (Project No. 055217); iv) the Ontario Ministry of Economic Development, Job Creation and Trade ORF-RE09-078 (Project Nos. 053970, 054345); and v) the Natural Sciences and Engineering Research Council of Canada (NSERC), Canada Research Chair (CRC) program Project No. 460788.

24e Ethanol-Water Solutions at Elevated Temperatures for Isolating Ultraclean Corn Stover Lignins

Bronson Lynn, Graham W. Tindall, Villő E. Bécsy-Jakab, David Hodge, Mark C. Thies

Corn is the most abundant crop in America, yet the early push for realizing the energy potential of the non-food portion of the plant via cellulosic bioethanol has largely stalled in recent years. The current high cost of production has kept it from becoming competitive with fossil fuels or even traditional modes of corn ethanol production. By extracting value from the lignin co-product, cellulosic bioethanol could help supplant the renewables market without affecting the supply of valuable feed corn.

The dispersity and complexity of lignin has long presented a challenge to valorization, though if an effective method for molecular weight fractionation can be employed, this could be turned into an opportunity. Considering one particular pair of applications, polyurethane foams and carbon fibers, we can demonstrate the effectiveness of our ALPHA fractionation and purification process by transforming raw bulk lignin into precursors for these different applications.

Taking raw alkaline pretreated corn stover lignin containing almost 11% sugar and 3% ash, we were able to isolate high molecular weight lignin fractions containing less than 0.05% sugar and 0.05% ash, while simultaneously producing lignin an order of magnitude lower in molecular weight that served as an effective precursor for polyurethane foams. All this was accomplished using only ethanol-water solutions at mild temperature conditions (60-75°C). This approach was also applied to a lignin derived from hybrid poplar trees, and yielded similar results. Due to the difference in properties between the corn stover and hybrid poplar lignin and the softwood lignin that has been the subject of this technique previously, changes had to be made to the ALPHA process to account for the abundance and properties of the present impurities. This new approach to ALPHA enabled this significant reduction in impurities, while presenting interesting insight into the molecular weight fractionation behavior of both the corn stover and hybrid poplar lignin.

This ALPHA processing scheme relies on shifting the composition of the ethanol-water-lignin system to alter the solution strength. First, to selectively remove the impurities before precipitating out the lignin of controlled molecular weights. What we observe is that the highest molecular weight lignin does not precipitate out of solution during the first decrease of solvent strength, but instead much later when the solvent has already become weaker. The contribution of chemical functionality to this behavior is being investigated by ^{31}P NMR, and will give insight into why this behavior is occurring and how we may be able to leverage it further.

Ultimately, the goal of creating high quality precursors for value added applications will come down to optimizing the properties. Through these adaptations to the ALPHA process, we have achieved some of the best corn-stover-based carbon fiber properties ever reported in literature as well as polyurethane foams that exceed ASTM standards even with very high lignin substitution rates.

24f Reactive Compatibilization of Biodegradable Blends from Phbv and Pbsa: Study on Effect of Chain Extender on the Mechanical, Thermal and Morphological Properties

Patricia Feijoo, Amar K. Mohanty, Arturo Rodriguez-Uribe, José Gámez-Pérez, Luis Cabedo, Manjusri Misra

Being aware of the global problem of plastic pollution, our society is claiming for new biodegradable materials to replace conventional polymers in short-life and single-use products like packaging, bags or cutlery. However, balancing their mechanical performance is required to increase their presence in the market. Brittleness of biodegradable poly-3-hydroxybutyrate-co-3-hydroxyvalerate (PHBV) was attempted to be decreased by melt blending with the flexible biopolyester poly(butylene succinate-co-butylene adipate) (PBSA). A epoxy-functionalized chain extender was used as a compatibilizing agent to enhance interaction between both immiscible polymers. The mechanical performance (tensile, flexural, impact) of three injection-moulded PHBV-PBSA compositions (70-30, 50-50 and 30-70 wt.%) in presence and absence of the chain extender was assessed. Morphology, rheology and crystallization behaviour were studied by scanning electron microscopy (SEM), frequency-sweep tests and dynamic scanning calorimetry (DSC). Crystallization of PHBV was hindered by the addition of the chain extender, which was reflected in the improvement of the mechanical performance of blends. In case of 50% and 70%-PBSA blends, homogeneity of results was increased while for 30%-PBSA blend elongation at break was 45% higher and flexural behaviour changed from brittle fracture to non-breakable. The addition of the epoxy-functionalized chain extender did not change the type of morphology developed by each blend nor the toughening mechanisms, so impact strength was barely affected in any blend composition. However, it reduced the size of dispersed phase domains due to a viscosity change. As higher the content of PHBV in the blend, the higher the effect of the chain extender in storage modulus and complex viscosity. Higher interaction between polymers led to increased mobility of PHBV chains and thus, improved processability.

Acknowledgements: The authors would like to thank the following for their financial support: i) Ontario Ministry of Agriculture, Food and Rural Affairs (OMAFRA)/University of Guelph-Bioeconomy for Industrial Uses Research Program (Project Nos. 030486, 030578; and ii) the Natural Sciences and Engineering Research Council of Canada (NSERC), Canada Research Chair (CRC) program Project No. 460788.

255a Fabrication of Waste Biomass-Derived Aerogel Using Zinc Chloride Salt Hydrate

Mairui Zhang, Yang Liao, Daniel Fougnier, Ivan Gitsov, Gyu Leem, Xuejun Pan, Jeong Jae Wie, Chang Geun Yoo

Aerogel is a highly porous material with unique properties including low density, thermal conductivity, and high water/oil retention capacity. It is mainly fabricated from synthesized polymers or cross-linked by additional chemical agents. With the increasing demand for green materials and sustainable processes, more attention has been drawn to developing aerogels directly from natural and renewable materials.

Cellulose is an abundant natural polymer and has been used for fabricating aerogels. It has advantages including its biocompatibility and non-toxicity. One of the cellulose-based aerogel fabrication approaches is the dissolution-regeneration process. However, one big challenge is the strong hydrogen bonds in and between cellulose chains impair the efficiency of most solvent systems.

Molten salt hydrate is an aqueous solution with a high salt concentration where the molar ratio of water to salt is near to the coordination number of the salt cation. Because of the similar properties of molten salt hydrates and ionic liquids, molten salt hydrates showed their capabilities to dissolve cellulose. Zinc chloride is a well-known cellulose swelling and dissolving agent with inexpensive price and low toxicity. Dissolved cellulose in zinc chloride salt hydrate has been reported to form hydrogel and films, while cross-linking is necessary to provide sufficient strength.

Waste biomass (i.e., end-of-life paper products) is a good resource for cellulose-based aerogel synthesis in terms of economic benefits. Waste fines and food packaging paperboards are typical unrecyclable biomass. Waste fines are rejected fibers in the paper recycling process because the short fibers are unable to provide sufficient strength for paper. Food packages have the issue as they contain different food residues. Hence, waste biomass is mostly landfilled or combusted. Also, calcium carbonate in these wastes can generate calcium ions, which are also well known as cross-linking agents for hydrogel formation, in the proposed salt hydrate system.

Herein, we report a facile route to fabricate cellulose aerogel directly from waste biomass using zinc chloride salt hydrate, where the role of calcium carbonate changes from impurities to cross-linkers to provide strength. The process is simple and under mild conditions. The fabricated aerogels were characterized using dynamic mechanical analysis (DMA), scanning electron microscope (SEM), thermal gravimetric analysis (TGA), and Fourier-transform infrared spectroscopy (FTIR). The aerogels showed high mechanical strength and lightweight with good thermal insulation performance. This study could pave a new way to valorize the waste biomass.

255b Evaluation of Physical and Morphological Alteration of Seaweed Char with Varying Activation Parameters and Corresponding Fate on Cationic Dye Adsorption

Cadianne Chambers, Toufiq Reza

Sargassum (brown seaweed) are free floating brown algae found on ocean shores in large quantities littering shorelines causing visual and olfactory nuisances affecting tourism and fishery industries for the affected areas. As Sargassum decomposes warnings have been issued to individuals with asthma or other respiratory illnesses to avoid the affected areas as the produced Hydrogen Sulfide may trigger breathing complications. As a result, the removal of Sargassum is imperative for sustaining the impacted industries. The use of Sargassum as a biomass source has proven to be a simple and wise decision due to the high organic carbon content and low cost. Superactivated hydrochar recently have gained substantial interest in adsorption applications such as gas storage, supercapacitors, dye, and toxin removal owing to the high porosity, and multimodal pore size distribution (macro-, meso-, micropores). Consequently, exploration of physical and chemical properties of the Sargassum derived Seaweed Char was undertaken. Synthesis involved hydrothermal carbonization (HTC) at temperatures 180, 220, and 260 °C for 30 min followed by chemical activation with KOH at activation temperature 800°C and varying KOH to hydrochar ratio (2:1, 3:1, and 4:1). Physical (Thermogravimetric Analysis, Ultimate Analysis), morphological (Brunauer-Emmet-Teller surface area, Scanning Electron Microscopy), and chemical characterization (Fourier Transform Infrared Spectroscopy, Boehm Titration) were performed to study the resulting Seaweed Char based on the varying synthesis conditions. From characterization results selected Seaweed Char were investigated for methylene blue dye adsorption application and as observed with increase in porosity of the Seaweed Char with results in an increase in maximum adsorption capacity.

255c Uncovering the Effect of Mechanochemical Pretreatment on Biocrude Yields and Chemical Mechanism of Lignocellulosic HTL

Heather LeClerc, Alex Maag, Geoffrey Tompsett, Michael T. Timko, Andrew R Teixeira

Lignocellulosic wastes have emerged as a popular feedstock for hydrothermal conversion to biofuels. They can be preferable to food waste and sewage sludge due to negligible nitrogen which has proven challenging in biocrude upgrading. Additionally, research has shown lignocellulosic feedstocks to result in low biocrude yields typically below 25%[1], decreasing their overall commercial suitability due to increased production costs. Previous research has estimated ball-milling to consume 5.6 MJ/kg bamboo feed[2], whereas biocrude typically results in higher heating values (HHV) between 30 – 40 MJ/kg, indicating the potential for pre-treatment to be energy-favorable.[3]

Mechanical pre-treatment has the potential to improve biocrude yields and process economics from lignocellulose-fed hydrothermal liquefaction. This work explores the effect of ball milling as a mechanical pre-treatment for lignocellulosic waste from two sources with varying cellulose and lignin contents. Preliminary results reveal that ball milling increases biocrude production from lignocellulosic waste by 95% compared to HTL at identical conditions with un-modified lignocellulosic material as seen in **Figure 1a**.

Figure 1b begins to uncover the underlying reasoning for the increase in biocrude yield. Micrograph analysis reveals that ball-milling results in a significant decrease in green waste particle size from 0.16 ± 0.09 mm to 0.05 ± 0.04 mm, corresponding to a decrease of nearly 70%. This does not fully account for the increase in biocrude yield, wherein **Figure 1b** indicates that ball milling decreases lignocellulose crystallinity, which also plays a role in increasing biocrude yield. To determine the specific contribution of particle size and crystallinity, un-modified lignocellulosic waste will be sieved into particle size ranges that mimic those found in ball-milled green waste and biocrude yields compared. Cellulose and lignin will undergo HTL at identical sieve fractions with and without ball milling to identify the specific contribution of crystallinity on biocrude yield.

This work aims the advance the potential for lignocellulosic hydrothermal liquefaction at a commercial level through an understanding of particle size and crystalline properties of the feedstock beneficial to increased biocrude yields.

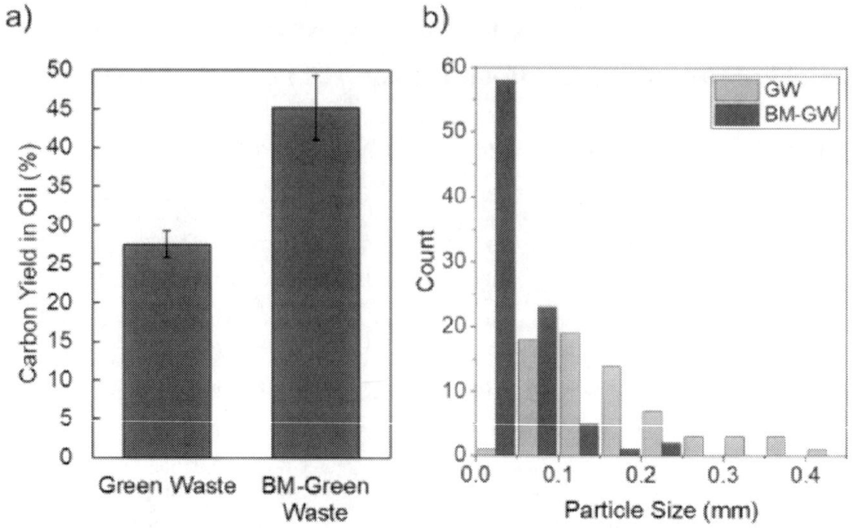

Figure 1. a) Biocrude yields for hydrothermal liquefaction of pure and ball-milled lignocellulosic (green) waste. HTL was performed at 300 °C for 60 minutes at 10% solids loading. **b)** Histogram of particle sizes for green waste and ball milled green waste feedstock. Particle sizes measured through the use of micrographs.

255d Structure-Property Relationships of Lignin-Thermoplastic Polyurethane Composites

Seoku Lee, Heeeun Lee, Mandeep Poonia, Kijoon Yang, Kwangho Kim, Chang Geun Yoo, Jeong Jae Wie

With the growing interest in eco-friendly polymer composites, natural-driven materials are highly illuminated for filler of the composites. Lignin is an aromatic macromolecule that consists of up to 30% of total biomass from woody materials and can be acquired in extensive quantities as a by-product of pulping and biofuel industries. Among various lignin utilization strategies, a polymer composite application was selected for this project. Polymer composites are polymer materials reinforced by various support materials which show enhanced mechanochemical properties. However, conventional composite materials highly rely on petroleum-based chemicals. To make this material more eco-friendly and economically feasible, we investigated biocomposite materials by incorporating lignin with ether-based thermoplastic polyurethane (TPU).

Prior to the melting mixing of lignin-TPU composites, the characteristics of lignins, extracted from softwood and hardwood via kraft and hot-water processing, were comprehensively analyzed by wet chemistry, nuclear magnetic resonance (NMR), gel permeation chromatography (GPC), thermogravimetric analysis (TGA), differential scanning calorimetry (DSC), and Fourier transform infrared (FTIR). Different lignins and TPU ratios were tested to find the optimal composite performance. The mechanical, thermal, and morphological properties of the produced lignin-TPU composites were analyzed by TGA, DSC, FTIR, scanning electron microscopy (SEM), and universal testing machine (UTM). The correlations between lignin properties and the technical performance (e.g., mechanical strength, thermal stability) of lignin-TPU composites were also investigated.

376a Exploiting Synergies of Anionic and Cationic Biomaterials for Oxygen and Water Vapor Barrier Properties

J Carson Meredith

There is strong interest in the introduction of renewable sources of packaging into global society. In particular, flexible barrier packaging and coatings are of interest due to their prevalence in the supply chain, e.g., at least 40% of all packaging is nonrenewable/nonrecycled flexible plastic. Forest-derived cellulose nanocrystals (CNCs) are a promising renewable source, and have been shown to be coatable as films with high oxygen barrier properties. Generally, CNCs are derived by acid hydrolysis processing that leaves pendant sulfate groups, imparting negative charge in aqueous media. Recent work in our group has shown that combining anionic CNCs with renewably sourced cationic nanofibers is a promising approach to developing manufacturable barrier films. Notably, chitin nanofibers (ChNFs) and nanowhiskers (ChNWs), which can be derived from renewable sources including crustacean food waste and fungi, are cationic and form complexes with CNCs that lead to dense barrier films. This talk describes recent innovations in optimizing the synergy between ChNWs and CNCs through careful tuning of the ChNW density and length. We find that optimal deacetylation of the ChNWs leads to formation of layered ChNW/CNC structures with oxygen permeability lower than poly(ethylene terephthalate) (PET). We also explore the use of thermal treatment and spray temperature to further improve the film structures. Finally, we will discuss an application of the bilayer films to PET film, to form PET/ChNW/CNC layered structures that have combined low oxygen permeability and low water vapor transport rate.

376b Improvements in Oil and Grease Resistance (OGR) Test Methodology for Waterborne Barrier Coatings

Allyson Marianelli

Paper-based food packaging is becoming more popular due to consumer demands for sustainable packaging options. Waterborne paper coatings that provide performance properties (i.e., resistance to oil and grease) not inherent to paper and board substrates offer improved sustainability profiles over earlier paper treatment options including fluorocarbon treatment and coating with extruded plastics. The continued development of new paper coating technologies requires re-evaluation of current test methods and development of new methods to ensure lab evaluations can serve as accurate predictors of real-world performance. This paper provides an overview of commonly used oil and grease resistance (OGR) test methods within the paper coatings industry, then describes improvements and developments made to two key methods: the 3M Kit test and an internally developed oil breakthrough test. The combined use of these adapted methods provides a more efficient testing workflow and a more complete understanding of the OGR performance of barrier coatings.

376c Value-added Biochemicals and Functional Biomaterials from Biowaste

Zhaohui Tong

Our society is facing long-term environmental challenges such as extreme climate change, resource and energy depletion, air pollution, food scarcity, and water and soil contaminations. To tackle these environmental and sustainable concerns and challenges, the use of green chemistry and engineering techniques to produce value-added bioproducts including platform chemicals, functional biomaterials, and biofuels from abundant renewable resources, especially low-cost waste becomes crucial. In this talk, I will present our studies on 1) fabrication of functional bio-based nanomaterials for mitigating water, soil, and food contaminants and 2) synthesis and production of chemical building blocks andamp; monomer precursors for sustainable polymer systems and hydrocarbon fuels. The applications of the bioproducts in food safety, agricultural, and environmental areas will be discussed. The key problems during the self-assembling, molecular design and synthesis, and catalytical valorization of the waste will also be elucidated and addressed using several recent works.

376d Conversion of Lignocellulosic Biomass to Value-Added Chemicals and Materials in Inorganic Ionic Liquid (molten salt hydrate)

Xuejun Pan

Molten salt hydrate (MSH) is a concentrated aqueous solution of an inorganic salt, in which the molar ratio of water to salt is close to the coordination number of the salt cation. Because MSH has similar properties to the ionic liquid, it is also called inorganic ionic liquid. The unique properties (such as low viscosity, high boiling point, low vapor pressure, enhanced acidity, and in particular the ability to dissolve cellulose) make it an efficient solvent and reaction medium for the processing and conversion biomass of lignocellulosic biomass. This presentation is to review our recent studies on the applications of LiBr molten salt hydrate (~61 wt% LiBr aqueous solution) in the fractionation of lignocellulose, hydrolysis and saccharification of cellulose and lignocellulosic biomass for sugar production, depolymerization and demethylation of lignin, quantitation of lignin and carbohydrates in biomass, synthesis of high-value chemicals and products (e.g., isomerization of glucose to fructose, furan-based chemicals, and prebiotic oligosaccharides) from simple sugars and biomass, and fabrication of cellulose nanocrystals, films, and aerogels from cellulose and biomass.

666a Hydrophobic and Water-resistant Lignocellulosic Packaging Materials Enabled by Metal Ion Modification

Xuefeng Zhang, R.M. Oshani Nayanathara

Lignocellulosic paper is an essential solution for preventing the increase in plastic packaging wastes; however, the inherent hydrophilicity and poor water resistance hinder its practical applications. Herein, we developed a facile metal-ion-modification approach for converting conventional hydrophilic, low-wet-strength lignocellulosic paper to hydrophobic and water-resistant paper with high wet strength. In this process, conventional paper is swelled with a dilute multivalent metal ion (e.g., Fe3+, Zr4+) solution, followed by drying to induce the coordination interactions between metal ions and lignocellulosic fibers. The resulting hydrophobic paper exhibits a water contact angle up to 140°, good wet tensile strength, and a low water absorptiveness of 10 g/m2. Hydrophobic paper is stable for long-term storage and solvent wash. This metal-ion-modification approach can potentially be applied in the wood pulping process for the scalable production of hydrophobic fibers and paper.

666b To Enhance Adhesion Property of Soy Protein Adhesive By Base-Solvents Modified Guayule Resin

Sarocha Pradyawong, Kimberly Ogden

Soy protein (SP) adhesives have shown a significant potential as renewable bio-based adhesives for wood products. However, the market value limited by its low water resistance property. Therefore, improving water resistance of SP adhesives will increase its market interest. Utilization of natural product as water resistance enhancer for SP adhesive provides a sustainable solution for healthy and environmentally friendly wooden products. Guayule resin, an attractive by-product of the guayule rubber extraction, is a good candidate for this purpose. The aim of this preliminary study was to investigate the water resistance of SP adhesive enhanced by using guayule resin. The resin was modified with NaOH in hexane, acetone, and water solutions, and subsequently mixed with SP adhesive at different concentrations. An addition of 7.5% resin to SP adhesive showed the highest elastic modulus representing the highest intermolecular interactions. The highest water resistance also observed at SP-blended with 7.5% resin. The wet adhesion strength significantly increased by 70.5% compared to pure SP adhesive. The study of denaturing behavior of SP suggested that guayule resin strengthened SP network by interacting with 7S and 11S subunits of SP. The changes in SP properties were due to interactions between guayule resin and protein's backbone and side chains. This study served as a strong motivator by enlightening research pertaining to the utilization of guayule resin as a bio-adhesives for wood industries.

666c Hydrolytically Stable Films from 3-Aminopropyl Triethoxysilan (APTES) Modified Cellulose Nanocrystals

Sadat Kamal Amit, Diego Gómez Maldonado, Maria Soledad Peresin, Virginia Davis

Commercially available sulfuric acid hydrolyzed cellulose nanocrystals (CNCs) are extensively explored as advanced functional materials. They possess high dispersibility in water due to having sulfate half ester groups on their surface, which enables easy processability. However, water dispersibility also prevents the use of CNC films in applications requiring hydrolytic stability. To address this issue, a scheme was developed to modify sulfated CNCs with 3-aminopropyl-triethoxy silane (APTES). The functionalization was confirmed with thermal gravimetric analysis, Fourier transform infrared spectroscopy, dynamic light scattering, elemental analysis, and energy dispersive spectroscopy (EDS). Studies of reaction stoichiometry and conditions showed that a 5% degree of substitution could be achieved without promoting significant aggregation. Tensile testing and nanoindentation studies of shear cast films showed no significant differences in the tensile strength and Young's moduli of films produced from the modified and unmodified CNC. However, APTES-CNC films had significantly more hydrolytic stability than unmodified CNC based on dissolution in water and mass loss studies. These results suggest that CNC-APTES modification can be promising for enhancing hydrolytic stability without compromising the inherent mechanical properties of CNCs.

AUTHOR INDEX

AIChE
120 Wall Street, FL 23
New York, NY 10005-4020

ISBN 978-1-7138-7886-5

9 781713 878865

CURRAN ASSOCIATES INC.
proceedings.com